THE EVOLUTION OF SEX

OTHER NOBEL CONFERENCE BOOKS AVAILABLE
FROM HARPER & ROW

NOBEL CONFERENCE XXIII

THE EVOLUTION OF SEX

John Maynard Smith
Lynn Margulis
Peter H. Raven

William Donald Hamilton
Sarah Blaffer Hrdy
Philip J. Hefner

Edited by Robert Bellig and George Stevens

1817

Harper & Row, Publishers, San Francisco

Cambridge, Hagerstown, New York, Philadelphia, Washington
London, Mexico City, São Paulo, Singapore, Sydney

FIRST EDITION

Library of Congress Cataloging-in-Publication Data

Nobel Conference (23rd : 1987 : Gustavus Adolphus College)
  The evolution of sex.

  Contents: Evolution of sex / John Maynard Smith—Sex / Lynn Margulis and Dorion Sagan—The meaning of flowers / Peter H. Raven—[etc.]
  1. Sex (Biology)—Congresses. 2. Evolution—Congresses.
I. Maynard Smith, John, 1920– II. Bellig, Robert.
III. Stevens, George. IV. Title.
QH481.N63 1987 575′,9 88-45156
ISBN 0-06-250291-3

---

88 89 90 91 92 HC 10 9 8 7 6 5 4 3 2 1

Contents

Contributors

John Maynard Smith is Professor Emeritus at the University of Sussex. He served as Professor of Biology at the University of Sussex from 1965 to 1985 and was the first Dean of the School of Biological Sciences (1966–1972). In 1977 he was elected as a Fellow of the Royal Society and foreign member of the American Academy of Arts and Sciences. He has written a number of scientific papers and books, including *The Theory of Evolution* (1958), *On Evolution* (1972), *The Evolution of Sex* (1978), and *On Being the Right Size, and other essays* (1985).

Lynn Margulis has been at Boston University since 1966, earning the rank of University Professor of Biology in 1986. She serves as associate managing editor of *BioSystems* (1983–) and is on the editorial boards of *Journal of Molecular Evolution* (1980–), *Origins of Life* (1981–), *Endocytobiosis and Cell Research* (1974–), and *Symbiosis* (1985–). From 1982 to 1986 she served as a NASA Advisory Council member. She was selected Fellow of the American Association for the Advancement of Science in 1975, and in 1983 she was elected to membership in the National Academy of Sciences. Dr. Margulis is the author of numerous articles, book chapters, and books including *Symbiosis in Cell Evolution* (1982), *Early Life* (1982), *Origins of Sex* (1986), and *Microcosms: Four Billion Years of Evolution from our Bacterial Ancestors* (1986).

Peter Raven has served as director of the Missouri Botanical Garden since 1971. He is also Engelmann Professor of Biology at Washington University in Saint Louis, Adjunct Professor of Biology at St. Louis University, and Adjunct Professor of Biology at the University of Missouri—St. Louis. He is chairman of the National Museum Services Board and a member of the U.S. National Academy of Sciences. Dr. Raven served on the editorial board of *Memoirs of the New York Botanical Garden* from 1966 to 1984 and currently works with the editorial board of *Evolutionary Theory* (1975–). He was winner of the International Prize in Biology presented by the Government of Japan in 1986. His books include *Coevolution of Animals and Plants* (1975), *Biology of Plants* (fourth edition, 1986), and *Biology* (1986).

William D. Hamilton has been a Royal Society Research Professor in the Department of Zoology and a Fellow of New College at Oxford University since 1984. Prior to that he was Professor of Evolutionary Biology at the Museum of Zoology and Division of Biological Sciences, Michigan University (1978–84), and was Lecturer in Genetics at the Imperial College, London University, from 1964 to 1977. Hamilton was elected a foreign member of the American Academy of Arts and Sciences in 1978. He is a contributor to journals including *Journal of Theoretical Biology, Science, Nature,* and *The American Naturalist.*

Sarah Blaffer Hrdy is Professor of Anthropology at the University of California, Davis. Her editorial work includes serving as consulting editor for the American Journal of Primatology (1980–) and *Primates* (Japan) since 1984, working on the editorial board for *Cultural Anthropology,* and being series editor of *Foundations of Human Behavior,* Aldine Publishing Co., since 1985. She is the author of works including *The Black–man of Zinacantan: A Central American Legend* (1972), *The Woman that Never Evolved* (1981), *Infanticide: Comparative and Evolutionary Perspectives* (1984), numerous journal articles, and several films.

Philip J. Hefner is professor of Systematic Theology (1967–) and Director of Graduate Studies (1978–) at the Lutheran School of Theology at Chicago. He has been a representative of Lutheran Church in America to the International Lutheran Reformed Dialogue since 1985. He cochairs the Consultation on Theology and Science of the American Academy of Religion and was president of the Institute on Religion in an Age of Science from 1979 to 1981 and again from 1984 to 1987. In 1984, he was a member of the Theological Task Force of the Commission for the New Lutheran Church. Hefner serves as an associate editor of *Zygon: Journal of Religion and Science* and is author, coauthor, or editor of seventeen books including *The Promise of Teilhard* (1970), *Defining America: A Christian Critique of the American Dream* (1974), and *Science and Theology in the 20th Century* (1981), as well as over one hundred articles and book reviews in various journals.

Preface

The very nature and purpose of Nobel Conferences at Gustavus Adolphus College is to bring together the world's foremost scholars to address significant issues of our time. A look back at the titles and participants of previous conferences will quickly confirm that a tradition of excellence has been established. This year's conference is no exception. We are fortunate to have assembled the very finest intellects to address an issue that is large enough and significant enough to warrant the title Nobel XXIII.

The evolution of sex is no trivial matter. Sex is so pervasive as to be found in some of the most simple to the most complex organisms; from microbes to mammals, from fungi to vascular plants. So many aspects of sex have eluded our understanding that sex is considered to be one of biology's great enigmas. If we fail to understand sex, then we can never fully understand the multitude of adaptations dealt to us in the "evolutionary card game." If we are ever to achieve a thorough and satisfying vision of life, we must understand the hows and whys of sex. The breadth of the ramifications of sexual activities extends from somewhere around meiotic microtubules to somewhere beyond nurturing males.

What we have just described is, of course, not the sex of the news media—sexual abuse, sex offenders, safe sex, and so on. This is the sex of genetic mixing, sex as an evolutionary force, sex as a socializing force. Biologists have wrestled with the meaning and purpose of sex and in the process have developed their own terminology and ways of thinking about sex. We view our role as editors as similar to that of a cartographer who can guide you toward uncharted territories. The chapter introductions were written to focus attention on how each chapter can be tied to landmarks you recognize. We have also provided a sample of the discussions that followed each address, as well as questions for further discussion, to show you where the contributors are headed.

A glossary is provided as an appendix to this volume to help you sort through the jargon.

Many of you are familiar with the issues at hand; you may have discussed them, even published on them. Much of what you already know is a result, directly or indirectly, of the contributors to this volume. It is our sincere hope that by bringing together the thoughts of these scholars we can provide you with new insights, imaginative ideas, and provocative challenges. We hope that the contributions of the assembled authors provide new fuel for your discussions with colleagues or students, and that the chapters that follow give you the flavor of Nobel Conference XXIII, The Evolution of Sex.

Robert Bellig
George Stevens
Gustavus Adolphus College

THE EVOLUTION OF SEX

In this chapter, John Maynard Smith sets the stage for the chapters to follow by describing in broad outline the unresolved controversies and recent developments in the study of sex. Maynard Smith has made a career our of searching for evolutionary puzzles and attempting novel solutions. His interest in "game theory" and its widespread acceptance started as an attempt to explain the apparent selfless behavior of losers in sexual or competitive contests. For example, why don't bull moose fight to the death for female favors? Maynard Smith's game theory provides an elegant solution that weighs the costs and benefits to the participants in the contests.

Maynard Smith's interest in the evolution of sex was initially motivated by his dissatisfaction with "group selection" explanations for its maintenance. At first approximation, sex appears to be inherently disadvantageous for the individuals that engage in it. To see this, compare the reproductive output of a sexual female who has the capability to produce two offspring, with that of an asexual (parthenogenic) female with the same physiological potential. In the first generation, there is no difference in the reproductive output of the two females. Each has two offspring. In the second generation (the grandchildren), the difference in reproductive output becomes apparent. In the sexual line, 50 percent of the generation is male and are physiologically excluded from producing young on their own in the second generation. The parthenogenic line is not so constrained. Counting grandchildren, one finds that the sexual lineage bears only two offspring, while the parthenogenic clone produces four. The males may have children through other females, but the physiological limits imposed on the sexual line by the presence of males produces what has come to be called "the twofold disadvantage of sex." One is left with the conclusion that the short–term disadvantage of sex must be balanced by some long–term advantage (otherwise, natural selection would have eliminated sex long ago). Because of this, evolutionists have long invoked a group–selection–based explanation for sex.

Pay particular attention to the structure of Maynard Smith's arguments: his method of building an evolutionary argument will be echoed in later chapters. Although each author has his or her own definition for sex, Maynard Smith is careful to explicitly derive both narrow and broad definitions for sex. In later chapters, it is useful to return to Maynard Smith's definitions while sorting out the different levels of analysis.

The Editors

1. The Evolution of Sex

JOHN MAYNARD SMITH

We are so used, in our own lives, to associating the ideas of sex and reproduction that we are in danger of forgetting that, at the cellular level, the two processes are precise opposites. Reproduction is the division of one cell to form two; the essence of the sexual process is that two cells fuse to form one. In fact, we can accept a broad or a narrow definition of sex. According to the broad definition, sex includes all those processes whereby genetic material from different ancestors is brought together in a single descendent. Such a definition would include processes such as bacterial conjugation and transformation. It has the disadvantage that it does not distinguish between sex and infection—for example, by virus or plasmid. But it could be argued that this is no great disadvantage, because in their earliest beginnings, sexual and infection processes were indeed not distinguishable. A narrow definition would confine the word sex to the process we observe in eukaryotes, which involves a regular haploid–diploid cycle, meiosis, and syngamy.

Whichever definition we adopt, we run into difficulties as soon as we ask questions about evolution. If we ask, "Why did the heart evolve?" we will answer in terms of the advantage that a heart confers in ensuring the survival of the individual organism. If we ask, "Why did the placenta evolve?" we cannot say that it contributes to the survival of the individual female, but it does contribute to the survival of her offspring. Most explanations in evolutionary biology are like this. We account for the evolution of an organ in terms of its contribution to the survival or reproduction of individual organisms. This is because Darwin's theory of evolution by natural selection predicts that organisms, because they have the properties of multiplication, variation, and heredity, will evolve traits ensuring their survival and reproduction. But to whose advantage is sex? The fusion of two cells is the opposite of reproduction. In higher eukaryotes, with separate males and females, things are even worse. A gene causing females to be parthenogenetic, producing only daughters like themselves, would double in frequency in every generation.

We cannot readily explain sex, therefore, by pointing to an advantage it confers on the individual. What, then, are the entities that are benefited by sex, and what is the nature of the benefit? Ultimately, I agree with Williams[1] and Dawkins[2] that the entities that benefit are the genes that cause sexual processes to take place. But I do not think there is any simple answer to the question of what benefit they obtain. Indeed, the answers may be different for eukaryotic and prokaryotic sex: that is, for its narrow and broad senses.

As an introduction to the difficulty, consider figure 1. I suppose that two viruses have infected a bacterial cell. Recombination takes place between them, mediated by a gene R on the bacterial chromosome, and gives rise to a new type of virus of particularly high fitness. But gene R gains no benefit from the recombination it has caused: it dies with the bacterial cell. Clearly an event of this kind does not confer a selective advantage on a gene such as R. Now there are many genes in prokaryotes that help to mediate

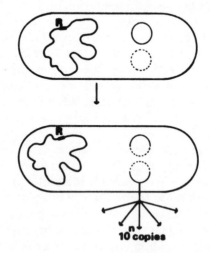

Figure 1. Recombination between two viruses infecting a bacterium.

recombination. I think it is now generally agreed that these genes have evolved, in part at least, because of the role they play in DNA repair. This first became clear when it was observed that rec⁻ bacteria (that is, bacteria with an impaired ability to recombine) are also particularly sensitive to ionizing radiation (the kind that damages DNA). The connection exists because the same molecular processes of cutting and splicing are involved in both DNA repair and in recombination.

I suspect that this may not be the whole story. There are, in prokaryotes, a number of processes whereby DNA is rearranged (for example, transposition, transformation, plasmid infection, and conjugation). Their evolutionary significance is profound, but I must leave discussion of that to others with a greater knowledge of microbial natural history than I have. My main topic is eukaryotic sex—sex in the narrow sense. There are, however, certain morals to be drawn from a study of prokaryotes that may be relevant to eukaryotes. The first, and most obvious, is that many of the genes involved in meiosis (particularly those concerned with recombination) originated as DNA repair enzymes in prokaryotes. The second is that DNA repair may also be relevant to the evolution of eukaryotic sex. The third is that, as in the example in figure 1, we are ultimately concerned with the survival and spread of individual genes.

As far as eukaryotic sex is concerned, perhaps the predominate view is that it owes its existence to the advantages it confers, not on the individual, but on the population as a whole. For reasons I will outline, I think there is something to be said for this argument. I must admit that this is somewhat ironic, since I first became interested in the evolution of sex because I had become convinced that "group selection" thinking—the idea that traits evolve because they benefit the group and not the individual—was mistaken, and its widespread, if tacit, acceptance was having disastrous effects on evolutionary theory. Sex was challenging because, as Fisher[3] had suggested, it was the one trait for which a group selection explanation was feasible. I will therefore first outline the argument that seems to favor a group selection explanation. I will then explain why group selection is more plausible in this context than in others, and then indicate why, despite its plausibility, it cannot be the whole story.

The strongest argument favoring an explanation of sex in terms of group advantage is a taxonomic one. Sex is the characteristic mode of reproduction among multicellular eukaryotes. Parthenogenesis is widespread in most animal and plant taxa, but its taxonomic distribution is spotty. That is to say, parthenogenetic varieties or species crop up in many taxa; but it is rare to find a whole genus, let alone any higher taxon, that is wholly parthenogenetic. (I exclude groups like the aphids and cladocerans, which are cyclically parthenogenetic: they still retain the long–term advantage of sex.) This is precisely the pattern one would expect if parthenogenetic varieties arise from time to time, and succeed in the short run, but are in the long run condemned to extinction in competition with their sexual relatives. There are exceptions to the taxonomic pattern: the Bdelloid rotifers, a whole order without males, is the most challenging. It is also important that among protists there are a number of taxonomic groups in which sexual processes have never been described. I do not know whether these groups are primitively asexual, or whether they are descended from sexual ancestors.

The taxonomic pattern, then, suggests that parthenogenetic varieties may be successful in the short run, but are in the long run condemned to extinction. The reason for the short–term success is not hard to see. Parthenogenesis confers a two–fold advantage. This advantage can be seen as the advantage of not producing males; or, at a "gene's–eye" level, as the advantage of suppressing female meiosis, and thus avoiding the 50 percent chance of being discarded in the polar body. It is also easy to see long–term disadvantages to parthenogenesis. For reasons illustrated in figure 2, sexual populations will evolve more rapidly than asexual ones. In a changing world, therefore, the parthenogens will lose out. Muller[4] suggested an alternative reason why parthenogens might suffer a long–term disadvantage. He showed that, in the absence of recombination, slightly deleterious mutations would accumulate by a kind of "rachet" process.[5]

Thus we have taxonomic evidence that group selection is relevant, and we can think of good reasons why parthenogenesis should have a short–term advantage, but a long–term disadvantage. Why, then, have I been reluctant to accept the group selection explanation? Essentially because, if individual selection is act-

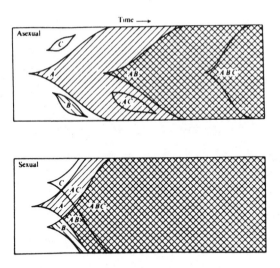

Figure 2. Evolution in asexual and sexual populations. The vertical scale represents the portion of the total population that carries a given allele. In the asexual population, novel mutations (for example C) are not as easily incorporated into the genetic material of successful lineages (for example A) as they would be were the population sexual. This leads to the loss of some flexibility in the evolution of asexual forms.

ing in one direction, and group selection in the other, individual selection is going to win. This is particularly true when the strength of individual selection is a factor of two in each generation. It is no good saying that group selection will favor sex in the long run if individual selection has made all populations asexual in the short run.

There is, however, a special reason why group selection may be effective in this case. This is that the origin of a parthenogenetic variety with a twofold advantage may be a rare event. If, for many millions of years, a population has been reproducing sexually, it may be very difficult to abandon the sexual process.

The nature of these "sexual hangups" can be best understood from some examples. First, consider the case of *Drosphila*.[6] Unmated females of many *Drosophila* species lay an occasional egg that develops without fertilization. However, to establish a successful parthenogenetic strain by selection in the laboratory, one must overcome two barriers: unmated females lay very few eggs, and very few unfertilized eggs start development. Even when these two difficulties are overcome, most of the eggs that do start development fail to hatch. This is because the eggs undergo meiosis and then restore the diploid state by fusing an egg and polar body (a process called *automixis*). Such restored diploids are homozygous at many loci, often including some that carry deleterious recessive genes: hence the failure to hatch. Despite all of the difficulties, there is one species, *D. mangabeiri*, that is parthenogenetic in the wild, although even in this species many eggs fail to hatch.

The same difficulties, and some others, are found in vertebrates. A number of fish and amphibia are parthenogenetic, but none has overcome the need for the egg to be penetrated by a sperm in order to initiate development. The parthenogenetic femals must mate with a male of another species, who provides sperm that trigger development (this behavior is called *pseudogamy*). The parthenogens can therefore only exist sympatrically with a related sexual species. Usually the chromosomes of the sperm are discarded. In some cases (some *Rana* and *Poeciliopsis*, the male's chromosomes survive in the cells of the hybrid, but are discarded in their germ line (this ploy is called *hybridogenesis*). Thus the animals are somatically sexual hybrids between two species, but only the female–derived genome is transmitted to future generations.

Lizards apparently have no need for a sperm to trigger development (or apparently no need for the sperm to provide a centriole). In both *Lacerta* and *Cnemidophorus*, there are species consisting wholly of females that do not mate. They have also solved the problem of genetic homozygosity through endomitosis. In this process, meiosis is proceeded by an extra round of chromosome replication, giving rise to a tetraploid cell. In meiosis, sister strands (that is, the two identical copies just produced by division) pair. The genetic result is to produce four diploid "gametes," each genetically identical to the original maternal cell. Thus *Cnemidophorus uniparens*, for example, consists of a single clone, within which skin grafts can be exchanged; but nevertheless it is a highly heterozygotic clone.

No wild birds or mammals are parthenogenetic. Parthenogenetic strains of chickens and turkeys exist in captivity; but they are highly infertile, mainly because their parthenogenesis is automictic (that is, they engage in the same form of selfing used by *D. mangabeiri*), leading to homozygosity. There is no reliable report of parthenogenesis in a mammal. This has been something of a puzzle, but a possible reason has recently emerged. It seems that, at some gene loci, the genes derived from the father and mother are differentially active in particular tissues. Since both activities are needed, every mammal must have a father and a mother.

The point of these examples is to show that organisms that have been sexual for a long time do not readily produce successful parthenogens. So, if the origin of new parthenogens is sufficiently rare, it is not impossible that group selection—the long–term elimination of parthenogenetic populations—can help to explain the presence of sex.

However, this argument must not be pushed too far. There are successful parthenogens. Most of them have suppressed meiosis altogether, or have meiosis preceded by endomitosis, as in the lizard example just cited: in either case, a female produces offspring genetically identical to herself. In animals, the cyclical parthenogens (aphids and cladocerans) appear to have suppressed meiosis. The same is true of the many parthenogenetic plants. The strongest argument against the group selection explanation is the "balance" argument of Williams.[7] He points out that it is not uncommon to find parthenogens coexisting, in the same region, with their sexual relatives. Since they appear not to be replacing the sexuals, even in the short run, Williams argues that there must be some short–term advantage to sex, and this advantage must lie in the genetic variability generated by sex.

There are empirical difficulties in evaluating this argument. Consider, for example, the snail *Potamopyrgus antipodorum* from New Zealand. Natural populations have an excess of males. This excess varies from place to place, but is fairly stable at any one location. Although some females reproduce parthenogenetically, it is still not certain what is really happening. There are two possibilities. One is that the snail populations are a mixture of typically sexual males and females, and of obligate parthenogens that produce only parthenogenetic daughters. In this case, the asexuals may consist of one or a few clones. In an ecologically variable environment, the sexuals will have an advantage in some ecological niches, and will not be replaced altogether, despite their twofold reproduction disadvantage. Vrijenhoek[8] has called this the "frozen niche hypothesis": asexuals are confined to particular niches. He finds that in the fish *Poeciliopsis*, clones do differ physiologically and in their ecological adaptations, and that the proportion of the population that is sexual is reduced as the number of distinct clones in a stream increases. In *Poeciliopsis*, however, the parthenogens need to be mated, so they cannot wholly replace their sexual ancestors. Nevertheless, Vrijenhoek's data do suggest one way in which the balance argument could work.

Returning to *Potamopyrgus* snails, the second possibility is that all individual females are facultative parthenogens (that is, they produce some offspring parthenogenetically, and some sexually). If so, the frozen niche hypothesis won't work. Unless we are going to argue that the proportions of the two kinds of offspring are fixed, then there must be some short–term advantage to sex. Williams argues that the advantage lies in the genetic variation among the offspring of a single female. He uses the famous analogy that a parthenogenetic female is like someone who buys one hundred tickets in a lottery, and finds that they all have the same number. I have argued[9] that this model works only if the offspring of one female compete with another: if offspring are dispersed, then a parthenogenetic female resembles a person who buys one hundred tickets, all with the same number, but in one hundred different lotteries. However, if there is intense competition between sibs, the model can give rise to a twofold advantage to sex. Sib competition could hardly be relevant to the maintenance of sex in animals with planktonic larvae such as oysters, but it could well apply to a snail like *Potamopyrgus*.

One difficulty in evaluating the sib competition argument is that we do not know how widespread is facultative parthenogenesis. It is known to occur in some plants: for example, in the grass *Dichanthium aristatum*. In this species, individual plants produce haploid eggs that need fertilization in the normal way, as well as diploid embryos that are genetically identical to the parent.[10] It is important that such cases should be identified and studied.

At this point, another distinction becomes important: the distinction between plants (or animals) that product offspring asexually from a single cell: and those that reproduce vegetatively through stolons, bulbils, or other multicellular propagules. Many plants, of course, can produce both sexually and vegetatively. But there are two reasons why the balance argument may not be applicable in such cases. The first is that there is usually an ecological difference between dispersive, sexually produced seeds, and nondispersive vegetative propagules. It may be that sex is retained because of its association with dispersal, and not because of the genetic variability generated. The second reason is one that has not, so far as I know, been emphasized before: if reproduction is through a multicellular propagule and not a single cell, deleterious mutations will accumulate more rapidly, leading to a higher "mutational load." This may rule out indefinite reproduction by multicellular propagules.

At this point, it may be helpful to summarize the argument so far. It goes as follows:

1. The spotty taxonomic distribution of parthenogenetic varieties suggest that, although successful in the short run, they are condemned to extinction in the long run.
2. This short–term success is expected because of the twofold cost of sex.
3. The long–term failure of parthenogens can be explained by the greater evolutionary potential of sexual populations, and perhaps by the accumulation of deleterious genes in parthenogens ("Muller's ratchet").
4. The justification for using a group selection argument of this kind is that the origin of successful parthenogenetic varieties may be a very rare event because of "sexual hangups" of various kinds inherited from sexual ancestors.
5. Group selection cannot be the whole explanation, because the coexistence of sexual and asexual forms shows that, in some cases, there must be a short–term advantage to sex. This advantage has usually been thought to lie in the greater genetic variability of sexual populations (following the balance argument).
6. There are two possible types of coexistence. The first is between sexuals and obligately parthenogenetic clones. The failure of the parthenogens wholly to replace the sexuals may be explained by the limited range of sexual genotypes

(as proposed in the "frozen niche" hypothesis).

7. Alternatively, there may be facultative parthenogens. This suggests that there is an advantage to the genetically variable sexual offspring of a single female, compared to the genetically uniform asexual offspring. This requires that there be competition between sibs.

8. The balance argument does not apply if individuals reproduce sexually through seed, and vegetatively by a multicellular propagule. This is because of the ecological difference between the two modes of reproduction, and because of the accumulation of deleterious mutations if reproduction is through a multicellular propagule.

The problem is complex because of the difficulty of deciding on the relative importance of long–term group selection processes, and shorter–term advantages of sex. This difficulty, and a comparison with the prokaryotes, in which it seems certain that the need for DNA repair has been a major selective force, has led several people to suggest that the generation of variability is not the important factor responsible for the evolution of sex even in eukaryotes, and that some type of repair is the essential thing. I want now briefly to discuss some of these suggestions.

Bernstein and others[11] have emphasized the importance of double–strand repair. The essential point is that if both strands of a DNA molecule are damaged, repair requires the presence in the same cell of the homologous chromosome. At first sight, this would seem to be an explanation for the evolution of diploidy, but not of meiosis and syngamy. However, the argument goes as follows. Diploidy, it has been suggested, may have originated because of the advantages of genetic complementation. Two haploid cells, carrying deleterious mutations at different loci, would gain by fusion: in effect, this is the same mechanism as underlies hybrid vigor. (This suggestion, of course, is not new, and it may well be right.) The argument continues as follows. Given diploidy, double–strand repair can occur. But when it occurs, it causes a crossover between the damaged and undamaged chromosomes. As a consequence, at the next division cells may be homozygous for genes distal to the crossover point. In other words, the crossingover associated with DNA repair has destroyed the genetic heterozygosity that was gained by fusion. Sex (meiosis and syngamy) evolved, they suggest, to restore the lost heterozygosity.

The argument is ingenious. As I will explain below, there is one feature of it that I now think may be correct. There are, however, two reasons why I find the argument as a whole unacceptable. The first is that it rests on the belief that is is impossible for double–strand damage to occur without, at least sometimes, a crossover taking place. (If this were not so, diploids could repair damage without risk of becoming homozygous). Now there is no logical reason why this must be so. There are at least some cases (such as yeast mating type switching) in which repair–like processes do occur without causing crossing over. If Carpenter[12] is right, then gene conversion events not associated with crossing over, and crossover events, occur at different times during meiosis. But my real difficulty here is more of a gut feeling than a reasoned argument: I find it hard to accept that meiosis, fertilization, sexual dimorphism, bird song, the color of flowers, Romeo and Juliet, all evolved because a molecular mechanism of double–strand repair not involving crossing over did not. Somehow this explanation seems inadequate to explain the phenomenon.

Clearly, we should not rely on gut feelings in science. There is more direct evidence that asexual diploids do not necessarily become homozygous. Many diploid clones, reproducing with meiosis, or by the genetically equivalent process of endomitosis followed by meiosis, do retain heterozygosity. A striking example that has already been discussed is the highly heterozygous lizard "species" *Cnemidophorus uniparens*. Such populations suggest either that double–strand damage is less important than Bernstein and company suggest, or that it can be repaired without causing homozygosity. A similar conclusion follows from the fact that there is no crossing over in male *Drosophila*, and yet over 99 percent of their sperm can be genetically undamaged.

Two other similar "repair–type" explanations of sex have been proposed. Holliday[13] suggests that the function of pairing in meiosis is to correct, not DNA damage, but incorrect methylation patterns. Bengtsson[14] suggests that meiotic pairing affords an opportunity for biased gene conversion. His argument is as follows. Suppose that some types of mutations are commoner than their opposites—for example, that the deletion of a base is more common than an insertion. Since most mutations are harmful, selection would favor an opposite bias in gene conversion: that is, when two homologues differ by the presence of one base, it should be more common to correct the mismatch by adding a base to the deficient gene than by removing a base from the other.

I think that both these arguments, Holliday's and Bengtsson's, suffer from essentially the same difficulty as that of Bernstein's group. They are arguments for diploidy, but not for sex. However, they need not be wholly mistaken. Given that meiotic pairing has evolved, for other reasons, it does provide an opportunity for changes in methylation, and for biased gene conversion. In particular, Bengtsson is clearly right to argue that there will be selection for enzymes that cause gene conversion to be biased in the opposite direction to mutation.

I mentioned above that there is one feature of the Bernstein group's argument that I find convincing. This concerns the nature of meiosis itself. Why, they ask, is meiosis preceded by chromosome doubling? The question is a good one. If the function of meiosis is to produce haploid cells from diploid ones, it would be sufficient for pairing to take place between homologues, without previous chromosome replication, and for pairing to be followed by a single reduction division producing two haploid cells. If genetic crossing over is an additional function, this too could be achieved by recombination between homologues, again without need for premeiotic doubling. This question has worried me for many years. I was therefore fascinated to learn from Margulis and Sagan[15] that there are protists in which meiosis involves only a single reduction division, and no premeiotic doubling. It may be a secondary simplification, but it is tempting to see it as primitive. In either case, Bernstein and company have an obvious explanation for premeiotic doubling. As they point out, it provides each chromosome with a sister chromatid that can serve as a template in double–strand repair. One can accept this part of their argument without accepting the rest. It will be important to find out whether repair between sister strands does occur during the period, sometimes protracted, between premeiotic doubling and the first reduction division.

I want now to turn to a context in which we can more or less rule out group selection as a relevant force. This is the evolution of recombination rate in eukaryotes. The reason for ruling out group selection is that there is good evidence[16] that there is genetic variance for recombination rate within populations. Selection experiments (for example in *Drosophila,* mice, locusts, and lima beans) have been successful in altering the recombination rate. If there is genetic variance within populations for a trait, then the level will be determined by short–term selective forces.

I have recently reviewed[17] the evolution of recombination. Much of the argument is parallel to that above, concerning sex. I want to emphasize an additional point. If a trait such as size is influenced by genes at many loci, some of which are linked, there can be strong selection on genes that alter recombination rate.[18] This can favor either an increase or a decrease in rate. If selection is normalizing (favoring individuals of average size, at the expense of the very large or very small), recombination rate declines. If selection is directional, recombination rate increases. There is some empirical evidence that strong directional selection increases rate of recombination. Burt and Bell[19] compared "excess chiasma number" (that is number of chiasmata, minus the haploid number chromosomes—the logic being that one chiasma is needed to ensure disjunction[20]) in a number of mammals.[21] They found that the highest values occurred in domestic animals.

The finding that directional selection causes increased recombination rates is an aspect of a more general point. This is that any explanation of sex in terms of its role in generating variability will work only if organisms are often exposed to directional selection. The same point underlies the argument that parthenogens go extinct because they cannot evolve as rapidly as sexuals. If the environment was constant in time, and selection was for constancy rather than change, I do not think many organisms would be sexual.

This conclusion may seem paradoxical. In recent years, paleontologists have been emphasizing the importance of stasis—lack of morphological change—in the fossil record. Dramatic environmental changes do occur, but not as often as all that. In any case, the response of a species to an ice age is likely to be, not the evolution of cold tolerance, but a shift of geographical range. Can it really be true that species, much of the time, are exposed to directional selection?

The clue may lie in the observation[22] that sexual reproduction is most often lost in habitats with relatively few species, and retained in biologically rich habitats. The important aspects of the "environment" of any species are its competitors, its predators, its prey, and its parasites. As they change, it must change: and as it changes, they must change. This has been called the "Red Queen" hypothesis.[23] Recently, particular attention has been concentrated on the host–parasite interaction, for reasons to be given in Hamilton's contribution to this volume. Our best chance of explaining sex seems to lie with the Red Queen.

NOTES

1. Williams, G. C. 1966. *Adaptation and Natural Selection.* Princeton University Press: Princeton, NJ.
2. Dawkins, R. 1976. *The Selfish Gene.* Oxford University Press: Oxford, England.
3. Fisher, R. A. 1930. *The Genetical Theory of Natural Selection.* Oxford University Press: Oxford, England.
4. Muller, H. J. 1964. The relation of recombination to mutational advance. *Mutation Research* 1:2–9.
5. For a more detailed explanation, see Maynard Smith, J. 1978. *The Evolution of Sex.* Cambridge University Press: New York.
6. Templeton, A. R. 1983. Natural and experimental parthenogenesis. In: *Genetics and Biology of Drosophila.* M. Ashburner, H. L. Carson, and J. M. Thompson, eds. Volume 3C Academic Press: New York.
7. Williams, G. C. *Sex and Evolution.* Princeton University Press: Princeton, NJ.
8. Vrijenhoek, R. C. 1984. Ecological differentiation among clones: the frozen niche hypothesis. In: *Population Biology and Evolution.* K. Wohrmann and V. Oeschke, eds. Springer–Verlag: Berlin, Germany. 217–231.
9. Maynard Smith, J. 1976. A short term advantage for sex and recombination through sib–competition. *Journal of Theoretical Biology* 63:245–258.
10. Knox, P. B. 1967. Apomixis: seasonal and population differences in a grass. *Science* 157:325–326.
11. Bernstein, H., H. C. Byerly, F. A. Hopf, and R. E. Michod. 1985. Genetic damage, mutation, and the evolution of sex. *Science* 229:1277–1281.
12. Carpenter, A T. C. 1987. Gene conversion, recombination nodules and the initiation of meiotic synapsis. *BioEssays* 6:232–236.
13. Holliday, R. 1984. The biological significance of meiosis. In: *Controlling Events in Meiosis.* C. E. Evans and H. G. Dickinson, eds. Society for Experimental Biology Symposia 38. Cambridge University Press: Cambridge, England. 381–394.
14. Bengtsson, B. O. 1985. Biased gene conversion as the primary function of recombination. *Genetical Research, Cambridge* 47:77–80.
15. Margulis, L., and D. Sagan. 1986. *Origins of Sex.* Yale University Press: New Haven, CT.

16. See note 5, above.
17. Maynard Smith, J. 1987. The evolution of recombination. In: *The Evolution of Sex*. R. Michod and B. Levin, eds. Sinauer: New York.
18. Maynard Smith, J. 1980. Selection for recombination in a polygenic model. *Genetical Research, Cambridge* 35:269–277; and Maynard Smith, J. In press. Selection for recombination in a polygenic model; the mechanism. *Genetical Research, Cambridge*.
19. Burt, A., and G. Bell. 1987. Mammalian chiasma frequencies as a test of two theories of recombination. *Nature* 326:803–805.
20. One argument has been that chiasmata (and hence crossovers) exist to ensure proper disjunction in meiosis. I do not find this convincing. Meiosis without the formation of chiasmata has evolved several times (for example, male diptera, female lepidoptera). Chiasmata can be localized at the ends of chromosomes, so that disjunction is ensured without crossing over. If it were selectively advantageous, meiosis without crossing over could have evolved.
21. Only males were used in this comparison, because chiasmata are hard to see in oogenesis.
22. Levin, D. A. 1975. Pest pressure and recombination systems in plants. *The American Naturalist* 109:437–451. Glesener, R.R., and D. Tilman. 1978. Sexuality and the components of environmental uncertainty. *The American Naturalist* 112:659–673. Bell, G. 1982. *The Masterpiece of Nature*. University of California Press: Berkeley, CA.
23. Van Valen, L. 1973. A new evolutionary law. *Evolutionary Theory 1:1–30*. Stenseth, N. C., and J. Maynard Smith. 1984. Coevolution in ecosystems: red queen evolution or stasis? *Evolution* 38:870–880.

PANEL DISCUSSION

HAMILTON: It seems to me that your explanation for the increased recombination rates observed in domesticated animals is not the only explanation. An alternative is that when animals are domesticated they are kept at much higher densities than they are found in the wild, and therefore their parasite problems go up proportionally. They need more recombination in order to fight those parasites. Have you any comment on that idea?

MAYNARD SMITH: No, I think I buy that as a perfectly plausible explanation. There is a possible third explanation that has worried me. Domesticated animals tend to be very inbred. This is something that inevitably happens to domesticated animals. But it turns out that if you just look at the effect of inbreeding by itself on chiasma frequency, it tends to reduce it. In reference to your point, however, I would be quite willing to buy the idea that recombination rates might be influenced by the high parasite loads in domestic animals.

MARGULIS: Following your line of reasoning that sex is a way of dealing with uncertainty, one should expect to find species that are facultatively sexual when the environment varies a great deal, but who go into an asexual mode when the environment stays very constant. Does one find that in nature?

MAYNARD SMITH: No, I do not know of any studies that suggest that kind of behavior. As a matter of fact, I can give you very few examples of organisms that are known to be facultatively parthenogenetic. In the great majority of animal cases, some individuals are obligately parthenogenetic, while other individuals are obligately sexual. We badly need to identify and study the ones that are facultative. I entirely take your point. There ought to be such cases; but I think the failure to find them is really a failure to look for them, not that they are not there.

AUDIENCE QUESTION: Have we identified a gene for sex? Do we know what we are talking about when we discuss the genetic changes responsible for the evolution of sex?

MAYNARD SMITH: Well, it is easier to talk about whether we can identify genes for not being sexual in the sense that if you haven't got sexual reproduction, genetic analysis is rather difficult. A typical situation is finding parthenogenetic varieties arising within a sexual population. In dandelions, for example, it is curious that many of the asexual parthenogens actually produce pollen, but they produce seeds which develop without fertilization. Their pollen can pollinate other plants; and when that happens, the pollen progeny are parthenogenetic as well. In other words, parthenogenesis is a dominant trait. As far as I know, nobody has done a detailed analysis of that, but it is clear that there is at least one dominant factor that induces parthenogenesis.

MARGULIS: Let me expand upon that question. You need to think about a gene for sex as being either capital S or small s, comparable to a gene for intelligence, capital I or small i. It would be very easy to make a tiny amino acid substitution (for example, in tryptophan metabolism) and lose everything associated with intelligence. On the other hand, the gaining of a trait by a single amino acid substitution is patently ridiculous. In the same sense, you can lose your sexual processes much easier than gaining them. When you talk about a gene for sex, you are talking about hundreds of closely coupled physiological and genetic features that give you the phenomenon. I would say to talk about a single gene for sex is nefarious.

QUESTIONS FOR FURTHER DISCUSSION

1. In discussing the evolution of sex, why is an understanding of the genetic basis of sex important?
2. Why is Maynard Smith so uncomfortable with the group selection explanation for the maintenance of sex?
3. Would Maynard Smith argue that sex arose because of the long–term advantage it would provide to the populations that engage in it, or does the group selection argument just provide a basis for understanding the maintenance of sex?
4. Maynard Smith has been correct often enough that his "gut feelings" are not to be easily ignored. What is his point about Romeo and Juliet?
5. Why is the observation that recombination rates increase with directional selection given such strong emphasis in Maynard Smith's address?

6. Are experimental methods of inquiry available to study the evolution of sex, or must all arguments be based on the interpretation of "natural experiments" already completed as the outcome of biogeographic and ecological differences between existent species?

Lynn Margulis and Dorion Sagan challenge the conventional view that sex can be understood by studying animals such as *Drosophila*, marine worms, fish, or humans. They point toward a subvisible world, appearing earlier in the fossil record, whose inhabitants have bizarre sex lives and strange ecological interactions. Their method is to look for the oddities that defy adaptationist interpretations; that is, phenomena that are imperfect in their present form and appear to have evolved in response to conflicting (often historical) pressures.

Look for a different level of analysis by Margulis and Sagan as compared to that of Maynard Smith. The microbiologist's perspective is best illustrated by the approach of Margulis in her answer to the question of whether or not a gene for sex exists (presented in the Panel Discussion section following Maynard Smith's contribution). While both Maynard Smith and Margulis would agree in their definitions of sex, they clearly are asking different kinds of questions as to its evolution.

The Editors

2. Sex: The Cannibalistic Legacy of Primordial Androgynes

LYNN MARGULIS AND DORION SAGAN*

We have written two books,[1] two chapters,[2] and two magazine articles[3] on the subject of the evolution of sex, and because of this we are considered knowledgeable. A hasty judgment: the complexity of the issue of sex renders us all equally naive. If we can claim special insight, it might flow from our familiarity with a small sample of the sexual oddities of the microcosm, the subvisible world with its *biota incognita*. We assume many modern microbes represent variations on the theme of simpler sex in earlier times. As voyeurs of the swarming world of life that can be seen only with one or another microscopic devices, we take our clues from organisms that really are neither animal nor plant, but ancestral to both.

Throughout history, people forced microbes into either one or the other of two great conceptual categories; every organism was deemed either to be a plant—a member of Kingdom Plantae—or an animal—a member of the Kingdom Animalia. But the inhabitants of the varied, wet world of microorganisms are neither animals nor plants; they don't fit in either kingdom. Spying at the tiny world that inhabits and surrounds us has led to the recognition of at least three other kingdoms of life. Life forms not plant or animal belong either to the Kingdom Monera (bacteria, also called the Prokaryotae), the Kingdom Fungi, or the Kingdom Protoctista (protoctists).[4] We emphasize these organisms of the protoctist kingdom; they appear to be crucial to any discussion of sexual beginnings. About 200,000 different species of extant protoctists are estimated, and from them the murky past of sex's

---

*Dorion Sagan contributed to this chapter, but was not a participant at the conference.

cellular beginnings can be reconstructed, shedding light on the sexual present in more complex sexual beings such as mammals.

All these nonanimals and nonplants (bacteria, protoctista, and fungi) live by rules and engage in sexual practices that are foreign to us. When compared with the overwhelming sexual complexity of the microcosm, human sexual differences become negligible and nearly identical to the attributes of all other four thousand species of mammals. We share with all other animals the same sort of sexuality, the same sort of "male–sperm–female–egg–fertilization–embryo–fetus" way of life. Because of the paucity of fundamental differences in animal sex, it is useless to study animals to understand the most ancient origins of the sex practices. Rather the oddities and peculiarities—the variations of the sexual theme—are in the bacteria, protoctists, and fungi, where the getting together of whole bodies began. The delicate sex lives of the microworld can illuminate our normally one–sided view of life, because we animals share a common evolutionary history with at least some of the bacteria and protoctista.

For the present purposes, we must state how we perceive the problem of the "origins of sex" and what possibly can be an acceptable solution to the problem. For the biologist, sex is far more—and in some sense much less—than human copulation. Sex is not reproduction. The biologist recognizes that sex, regardless of its peculiarities, must begin with at least two interacting individuals, one of which must be alive. Sex is a happening, a series of processes that occur over characteristic periods of time and that have generally predictable outcomes. Sex is never a single object or action.

We distinguish sex from reproduction now and describe both these complicated processes a little later on.

The sexual process is defined as *a series of related events which create a genetically new and distinctive live being, an individual who has more than a single parent.* Because human copulation occasionally results in fertilization of an egg and thus the conception of a new human being, our view of the fundamentals of sex is colored by the special relationship human sex has to reproduction. Sex has been a necessary step to ensure continuity through time of our *Homo spaiens* species. Most people (but not biologists) when asked, will assert that sex is the mating act, the process that can lead to human reproduction. Many professional biologists on the other hand will claim that sex is the process that ensures genetic variation in any natural population of animals, plants, or microbes. In defining sex as the process making genetically new individuals from more than a single parent (clearly separating sex from reproduction), we remove our focus from the first definition. Yet because many scholarly studies have shown the lack of a direct relation between two–parent sex and genetic variation, we must also reject the second statement—even though it is a standard dictum and still found in many texts and articles.

Reproduction is the increase in numbers of beings. At the beginning of an act of reproduction, if there are N individuals, at the end of the act there will be at least $N + 1$. But among the thirty million or so extant species of life on Earth, of which *Homo sapiens* is only one, it is easy to demonstrate many cases of reproduction in the complete absence of sex. Sex in the absence of reproduction is equally well known. So, although sex and reproduction are commonly associated, biologically, they are entirely separable. Furthermore, genetic variation, no matter how measured, is easily shown to be present in great abundance in organisms that have lost sex or seem never to have acquired it. By the same token, great uniformity can be seen in some organisms that regularly reproduce by sexual means. The fact that identical twins come from the asexual mitotic division of a single egg does not mean that sex is required to generate variation. Our views are tempered by the fact that, as people, we happen to belong to the class Mammalia, where sex and reproduction are tightly correlated. But in most of the rest of life there is no obligate association of sex and reproduction. A single stalked head of an *Aspergillus* mold, unmated and without benefit of any sex, can reproduce itself, making some ten thousand spores per minute for months.

Time and space preclude long descriptions of the putative origins and evolution of details of sexual processes. This kind of material can be found in more technical works.[5] Here we simply state our speculative conclusions, only hinting at how they are derived. Our major point is that complex mammalian sex lives are products of a cumulative evolution: ancient historical pathways become inextricably intertwined. With difficulty, the strands can be unwoven and our histories can be factored into their component parts, and, in principle, complications based on historical precedent can be revealed. Whereas certain aspects of our multifaceted sex lives extend back nearly to the origin of life itself (for example, recombination of our DNA molecules began over 3,500 million years ago), other aspects of our sexual existence, such as love poems and types of flirtation, evolved far more recently. The fundamental material aspects of sex (genetic, biochemical, and physiological) are still with us because they are directly related to the desperate strategies of survival of particular identifiable ancestors. Over the course of an extremely protracted evolution, events took place that constrained who we are and who we can be, events that led from our ancestors to ourselves.

No one can reconstruct with certainty the details of the more than 3.5 billion years of evolution of life, including the sex life of organisms on earth. But the general outlines may be accessible to verification. We couple the nineteenth–century method of Charles Darwin with clues taken from myriads of contemporary studies of live organisms. Darwin admonished us not to ignore the "oddities and peculiarities" of life as we see it today. It is by the analysis of such oddities that evolutionary history can be reconstructed.[6] Evolution is opportunistic and never foresighted; evolutionary happenstances and responses to exigencies accumulate. In studying oddities and peculiarities, biologists have concluded that the panda's "thumb" evolved later than and entirely separately from the thumb of our primate ancestors; that certain fish jaw bones became the ear bones in reptilian and mammalian descendants; that the bladder, the uterus, and the bowel (colon) were a single common organ (the cloaca) in our mammal–like reptilian ancestors. Oddities and peculiarities are explicable as historical records, not as committee recommendations or premeditated designs. In sum, Darwin recognized that oddities and peculiarities are legacies of ancestral life. As Stephen J. Gould puts it, "Darwin answers that we must look for imperfections and oddities, because any perfection in organic design or ecology obliterates the paths of history . . . This principle of imperfections became Darwin's most common guide . . . I like to call it the 'panda principle'" To apply Darwin's "panda principle" to the origin of sex, we must study the current sex lives, eating habits, and survival processes of an assortment of peculiar living microscopic forms. We parsimoniously assume that many micro-beings have retained lifestyles from far–earlier times. Knowing that the observations of *now* might have been the methods of *then*, we string together the observable events to make a coherent whole. Our main method is the analysis of sexual diversity, primarily as seen in the wildly disparate group of protoctist organisms, many of whose habits are astounding. Our analysis is restrained at the outset by two factors. First, our evolutionary reconstructions must be consistent with the fossil records of life on earth. Second, we must not invent lifestyles and sex acts in ancestors for which there are no precedents in organisms living today—that is, we cannot invent "missing links." On the contrary, we take as given the conservatism of the evolutionary process: we assume that we can find among extant bacteria, protoctists, and fungi (and sometimes the smaller animals) examples of all the steps we hypothesize for the origin of the convoluted sex practices in which we ourselves indulge. In this way scientific methods are used to recreate sex's vanished history.

FIRST SEX

The earliest sex, the transfer of genes in bacteria, is thought to have been a by–product of the need to protect against intense, damaging ultraviolet radiation from the sun. In the Archean Eon, beginning over 3,800 million years ago and ending 2,500 million years ago, the world's atmosphere contained no free oxygen. A consequence of the lack of oxygen was the absence of ozone. The ultraviolet light, along with visible radiation, surged from space to the surface of the Earth unfiltered by the ozone layer that now protects us.

The earliest sex, fundamentally DNA recombination, was itself probably a response by the earliest bacteria–like cells to that same life–threatening ultraviolet light. Survival became sex—by definition—as soon as a DNA molecule had two different parental sources. Today bacterial sex is still rampant.[8] Much bacterial sex depends on "drinking up" genes from other bacteria, from viruses, and even from isolated DNA molecules (see fig. 1). These genes are acquired from the surrounding water and incorporated into the recipient bacterium's body. In the cases of viruses, the genes may be protected by a protein coat, but many bacteria can take up genes as naked DNA. Such sex, if it were to occur in humans, would be astounding: it is as if a blue–eyed, curly–haired person could ingest a gene or two, say, for brown eyes or straight hair, and then pass genes on in a normal way to his or her children. Casual sex indeed: the way of the bacterium . . .

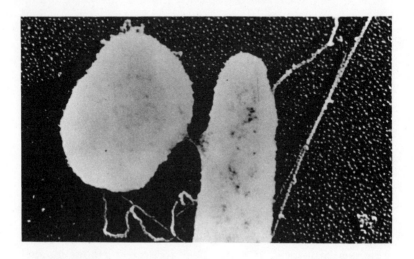

Figure 1. First sex, no reproduction. The bacterium on the right donates "his" DNA to "her" on the left.

Although in many cases a bacterium will wait and mate with another of its kind, the amount of genetic material it takes from its sexual partner is variable. The contribution from each of the bacterial parents is nearly always grossly unequal. The recipient bacterium may receive a huge injection of genes from its partner; on the other hand, it may receive hardly any. Furthermore, in its most simple manifestation, the sex life of a bacterium can depend essentially on DNA, a dead chemical. If you met someone who had a sex life analogous to that of the *Haemophilus* bacterium, you would consider it weird. Imagine a woman bypassing copulation and the vaginal acceptance of sperm. Instead, she would receive an injection or a potion of chemically purified genes (we might even imagine the genes coming from her long–dead husband). then, perhaps, nine months later she would bear a child who had some arbitrary number (that is, not necessarily 50 percent) of her genes. The rest of the child's genetic endowment would be derived from the donor DNA. An analogous sort of practice, the uptake, incorporation, and use of some other microbe's genes, occurs daily among the population of packed bacterial crowds in the soil, in your mouth, and in your intestine. Indeed, such freewheeling transfers, playing fast and loose with bacterial hereditary material, is an important mechanism in the upkeep of global ecology.

From direct evidence of the geological and microfossil record, we can infer that bacteria thrived in the Archean Eon, more than 3,000 million years ago. At that time, the Earth was inhabited only by bacteria, lodged in low–lying communities and threatened by brilliant light as soon as the sun rose for the five or so hours of the short day's light (Archean days were shorter because the earth turned more quickly on its axis). Ultraviolet sunlight is devastatingly dangerous to unshaded life: the DNA which comprises genes absorbs the light and chemically breaks down as a consequence. Many early bacterial cells must have died as their DNA was broken. Those that survived were the ones able to repair their DNA. To compensate for their own losses of DNA, they used DNA from various sources close at hand, including DNA from viruses and other bacteria. The survivors had engaged in the first sex: they contained genes from more than a single parent. In modern bacteria, nearly two dozen kinds of enzymes are specifically involved in patching, splicing, and rejoining DNA. Even today many of the DNA–patching processes can be induced by brief exposure to ultraviolet light. Such behavior seems to be an ancient legacy from the earliest days of sex. First sex had arrived, and it had come before gender, before sperm, before embryos, before even the evolution of cells with nuclei. Sex–the incorporation of some other individual's DNA to patch up one's own—was already a daily imperative for survival.

SECOND SEX

The second type of sex to evolve, the "meiotic sex" characteristic of animals and plants, generally requires the wholesale fusing of cells—first fusion of the cell membranes and then of the nuclei inside them. While meiotic sex of cells–with–nuclei started as a strategy of survival, it evolved much later in organisms very different from bacteria and against the backdrop of a chemically transformed planetary environment.

The second kind of sex is the kind that brings pollen to the female's ovaries in plants. The second sort of sex is what animal sperm and eggs indulge in. Typified by the fusing of swimming green algae (such as *Chlamydomonas*) or passive fungal balls or threads (such as yeast), meiotic sex has common attributes throughout the world of nucleated cells. The fusion of partners of the opposite sex is known as syngamy, or fertilization, conjugation, or mating. No matter the label, cells with a nucleus having a single set of chromosomes meet. The cells must recognize each other and then fuse. The entire nucleus (a spheroid filled with genes) of one mating partner fuses with the nucleus of the second mate, a fusion called *karyogamy*. The product of the sex act, the doubled nucleus, has two of everything—two nucleoli, two copies of the DNA, and extra nuclear proteins. A reduction of that doubleness must occur at some time in the life cycle of each of these newly mated beings, whether algae or fungal balls or threads, because the sleek, single form of the organism grows faster and more efficiently. Thus each doubled form must eventually compensate for its doubleness by halving its composition. The compensating step—the halving of the number of genes, the halving of the chromosome number in the nucleus—occurs through meiosis. Because all organisms that indulge in the doubling of fertilization have later to undergo the halving of meiosis, this kind of cell sex is called meiotic sex. Meiotic sex is found in the four nonbacteria groups of organisms, the protoctista, fungi, animals, and plants that taken together are called eukaryotes. Meoitic sex is so different in detail from the loose bacterial exchange of DNA that no scientist believes it developed directly from bacteria.

We suggest that meiotic sex is, like bacterial sex, a legacy of survival. Meiotic sex began in protists as a primeval act of desperation (see fig. 2) and became a prerequisite for the complex processes of histogenesis and organogenesis. By the time meiotic sex evolved in the Proterozoic Eon (from 2,500 until 570 million years ago), there was enough ozone in the air to mitigate against the destructive properties of ultraviolet light. Unlike bacterial sex, which arose in the face of lethal doses of sunlight, meiotic sex was driven less by physical threats than by biological ones: crises of drought and starvation.

Our thesis is that meiotic sex evolved as appeasement of the recurrent menace of hunger and thirst. In this view, sex and feeding were intimately and inextricably intertwined from the beginning. Sexual fusion began as a desperate primeval hunger and continued in the form of sustained cell doubleness. Cannibalism arose in the close original relationship between sex and death: organisms confronted with starvation had to eat, and eat each other, or die. Having eaten each other, death of a kind—a partial death—ensued, since parts of the cell (though not all of it) were digested. We see our complex sex lives today as permutations of these ancient protoctist activities.

Cell–fusion sex began in protoctist organisms with no mouths, no penises, no vaginas, not even stomachs or anuses. Cell–fusion or meiotic sex began in the intimate mingling of single–celled organisms and their food in their daily phagocytotic acts of feeding.

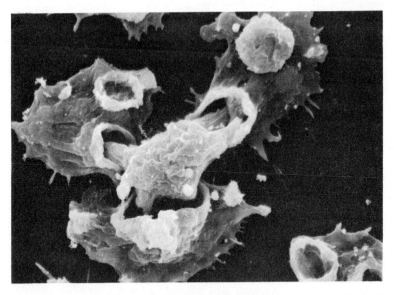

Figure 2. Second sex, desperate cannibals (protists). *Naeglaria fowlerii:* three cells attempting to cannibalize one victim. (Scanning electron micrograph courtesy of Dr. David John, Department of Microbiology, Oral Roberts Medical School, Tulsa, Oklahoma.)

When they began to starve or desiccate, they tried to eat, fused, and survived by successful digestion of their fellow cells. The fate of most cannibal victims was death. The fate of most of the doubled monsters was death. The fate of still other doubled monsters, temporarily with less surface area per unit volume, was beyond death: it was to develop mechanisms for relief of doubleness. The triumphant protists that successfully cannibalized their fellows are not necessarily relevant to our story. We are the descendants of those would–be victims, which survived by coexistence with their would–be cannibals rather than being digested by them. The Chinese ideogram of "danger–opportunity" for "crisis" tersely states the case (see fig. 3). In surviving the *dangers* of being eaten alive, our wily single–celled ancestors, more than a thousand million years ago, exercised cyclical *opportunity* of coexistence to survive the *crisis*. Populations of our ancestors were periodically eaten alive, but not digested by their fellows—season after season, year after year, and in some cases day after day. They were devoured, but resisted digestion. Our thesis is that sex is a legacy of that survival from hunger, a periodic imperative. The ancestral death threat was the danger that led to the opportunity of complex tissue and organ development.

If we humans still displayed the biology of the protoctists and fungi who periodically fuse, we would behave like the dinomasti-

Figure 3. Chinese ideogram: "danger–opportunity."

gotes or baker's yeast, or the chlamydomonads or desmids. How so? First of all we would have no gender; our bodies would be neuter. Only after we had released some precious drops of a genuine aphrodisiac, a protein–contining "sweat," would it be noticed that there were two kinds of us. One half of us would produce a substance identifying our mating type and about half would have the genetic wherewithal to respond to this single secretion. The mating–type substance is a powerful mate attractant to our genderless, but functionally detectable mate. Our round green bodies would exude one of the two types of attractants only when we were starving for nitrogen, the aphrodisiac serving as a signal of this starvation. After a protracted period of no meat, no tofu, no legumes—of nothing that contains nitrogen reserves—we become desperate. Sugars and starches lacking nitrogen would eventually become totally inadequate as food. Weakening, we would huddle with our equally distressed green globular friends. If the threat failed to pass we would fuse our body with that of any near neighbor—whomever we could find of the opposite mating type. We would take our partners in completely: we would become part of a doubled monster. As doubled monsters, now with just enough nitrogen to live, we wait through the famine or drought or winter storm until some new nitrogen source rescued us.

This is precisely what these protoctists do today. They starve, fuse, and then wait (see fig. 4). Which of the nitrogen–starved beings will survive to leave their own offspring? Those doomed to doubleness will die. Those that survive will be those that can emerge from doubleness to reconstruct a single form like the original being who swallowed its fellows whole.

Fusion, partial digestion, survival: these were protist prerequisites to the meiotic sex of protist descendants—animals, plants, and fungi. The reduction of doubleness is still a major part of the meiotic process, of course. But another part, an aspect of meiosis found in all animals, plants, and fungi, involves the pairing of chromosomes and various macromolecular syntheses. We believe that meiotic pairing, macromolecular syntheses, and other aspects of meiotic sex have become connected to complex embryogenesis, histogenesis, and other developmental processes of animals and plants.

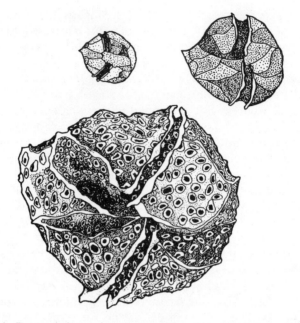

Figure 4. Protoctist's sex: hurry–up–and–wait inside the cyst. After fusing, dinomastigotes often form thick–walled cysts (*hystrichospheres*) like these. (Drawing by Sheila Manion–Artz.)

All of this implies that we, and all other animals and plants, are descendants of those who survived by returning to the simplicity of a single state. We are descendants of would–be cannibals periodically relieved. Today, because so many animals and plants have reverted to the one–parent, nonsexual state, we know that the several complex processes comprising meiosis (that is, reduction of chromosome number, chromosome pairing, and so forth) are separable from each other in time and space. Meiosis and fertilization in their glorious complexity are protistan legacies that later became ritualized (that is, fixed in animal and plant development).

Extreme environmental changes that force protist cells to fuse for survival occurs today in soil, on river banks, in drying pools, and in tidal flats. Many contemporary protists, not in any way our direct ancestors, will fuse when survival is threatened, but not necessarily even by twos. Ciliates like *Sorogena*, like *Acrasea*, and like amoebomastigotes, have orgiastic sex lives. Every time conditions around them become intolerable, they fuse by the tens of hundreds or even thousands. They swarm and throb, excitedly recognizing each other. They then engage in a massive orgy that is an analog of fertilization. Many cells fuse to make—relatively speaking—a giant being. Mistaking food for sex and sex for food, they merge into a moving mass that becomes much more than a doubled or tripled monster. In this evolutionary sideshow, only those capable of resolving or undoing the state of multiplicity and monstrosity in which they are embedded and of returning to a single body have survived until today.

Because the sex act varies so profoundly among protoctists, animals, fungi, and plants, we infer that meiotic sex evolved many times. The independent ancestry of sexually reproducing organisms can be concluded from the observation of the vast difference in the details of the return of each generation to an "uncoupled" state. In people, this return via meiosis to the single state is inconspicuous; it is very fleeting. The various meiotic processes occur to make eggs in women and sperm in men. (In women, meiosis occurs in the ovaries of a girl fetus; in men, it occurs in the testes of the postadolescent a few days before mature sperm are ejaculated.) The sperm and eggs of people are reminiscent of the ancient microbes ancestral to the entire animal lineage. They never survive very long (hours or days at most), but soon fuse to become a doubled being, or they die. The doubled being, the embryonic product of the fertile egg, then grows by cell division. The embryo gives rise to the fetus, which in turn becomes the infant, the child, the adolescent, and the adult. Each human cell at all of these later stages is doubled; each bears two sets of genes, the ancient imprint of our doubled ancestry.

Long before the recognition of the evolution or the mutability of species there were concepts, or rather myths, of sexual doubleness. Even today they are worth considering, as such myths embody and encapsulate in an intuitive and highly pictorial way much of the foregoing information. Nonetheless, myths of sexual doubleness were arrived at through introspection, not by natural history observation. From the Dogon peoples of West Africa, to the Babylonians and Maori, creation has been pictured as a perfect androgynous union between Mother Earth and Father Sky, engaged in a form of permanent copulation. In some myths the arrangement becomes disrupted, with the world parents separating to make room for humankind. Indeed, that these versions of a primordial bisexual union are so widespread in religious traditions and creation myths of different cultures suggests that this is an idea which, in one form or another, has often presented itself to human consciousness.

Perhaps the most eloquent expression of human doubleness arrives through the lips of Aristophanes in Plato's Symposium, the depiction of the banquet at which each guest offers an opinion on the meaning of eros or sexual love. During his turn, Aristophanes refers to the primordial form of humanity, in which men and women assumed a doubled state as Androgynes (see fig. 5). In this state, our ancestors appear as interlocked spheres, with two sets of

Figure 5. Androgyne. (Drawing by J. S. Alexander.)

limbs. Stronger than modern mortals, they moved about by rolling. Indeed, as with the Titans and Giants of Greek mythology, the primordial Androgynes were so powerful that the Gods began to worry. Zeus, to ward off the danger, split apart the primordial sphere into separate halves. Since then, the separate halves have been wandering the earth looking for their missing counterparts, and joining, if briefly, in attempts at reunion.

In an evolutionary update of this myth, we may think of the protoctists as real primordial spheres. However tiny, our ancestors were monstrous doubles, combined spheres more powerful than their single counterparts in extracting nitrogen, in surviving without water or food. They were not split through an outside agency, however, but naturally in the process of cell division. Nonetheless, they bear a striking resemblance to the Androgynes of Greek myth. Our bodies have retained the urgency of their protistan ancestors: our cells too must come together and fuse. The fusion urge results not merely from whim and the need to reproduce. It is a primordial legacy requiring us to reestablish the androgynous state that preceded our existence.[9]

NOTES

1. Margulis, L., and D. Sagan. 1986. *Origins of Sex*. Yale University Press: New Haven, CT. Sagan, D., and L. Margulis. *Mystery Dance: An Inquiry into Sexuality*. Summit Books: New York.

2. Margulis, L., and D. Sagan. 1984. Evolutionary origins of sex. In: *Oxford Surveys in Evolutionary Biology*. R. Dawkins and M. Ridley, eds. Volume 1. Oxford University Press: New Haven, CT. 16–47. Margulis, L., D. Sagan, and L. Olendzenski. 1985. What is sex? In: *Origin and Evolution of Sex*. H. O. Halvorson and A. Monroy, eds. Alan R. Liss: New York. 69–85.

3. Sagan, D., and L. Margulis. 1985. The riddle of sex. *The Science Teacher* 52:16–22. Sagan, D., and L. Margulis. 1987. Cannibal's relief: On the origins of sex. *New Scientist* 115:36–40.

4. Margulis, L., and K. V. Schwartz. 1987. *Five Kingdoms—An Illustrated Guide to the Phyla of Life on Earth*. 2d ed. W. H. Freeman Publishing: New York. Bacteria, the smallest of organisms, are defined by the lack of a membrane–bounded nucleus in their cells. Many are multicellular, but none has tissues or organs. The sex lives of such microorganisms and their descendants are bizarre. The Kingdom of Fungi comprises some 100,000 kinds of beings made of cells with nuclei, and includes molds, mushrooms, and puffballs. All fungi grow from spores, none from embryos. Most often when fungi have sex, two threads (for example, in molds) or two little spheres (for example, in yeast) fuse. The superimposition of any idea of "maleness" or "femaleness" on them is artificial, since—even though they are of complementary or opposite mating types—the two mating threads or spheres look identical. One might say, then, that fungi

have sex but lack gender. Protoctista is the name given to all organisms excluded from the other kingdoms. All protoctista are composed of cells containing nuclei. All protoctists grow from an embryo. Some protoctists, such as the giant kelps and other brown seaweeds, are well known and easily noticed. Others, like *Nanoclorum*, are so tiny that very, very few botanists have encountered them. The most familiar large groups of protoctists are the algae, protozoa, water molds, and slime molds.

5. For example, see Margulis and Sagan, *Origins of Sex*, note 1.

6. Gould, S. J. 1986. Evolution and the triumph of homology, or why history matters. American Scientist 74:60–69.

7. Ibid.

8. Sonea, S., and M. Panisset. 1983. *A New Bacteriology.* Jones and Bartlett Publishers: Boston, MA.

9. We thank Gail Fleischaker, Lorraine Olendzenski, and Simon Robson for helpful comments on this manuscript. We thank J. Steven Alexander for the drawing in figure 5. We are grateful to Madeleine Sunley and John Kearney for manuscript preparation and to NASA Life Sciences (NGR–004–025), the Lounsbery Foundation, and the Boston University Graduate School for support.

PANEL DISCUSSION

AUDIENCE QUESTION: At what point did sex become a nonsurvival act?

MARGULIS: That's a fascinating question. The problem is that sex is so tightly connected to what an animal is. Nearly all animals, and certainly all mammals, come from a sex act. Mammals then go through embryogenesis, gastrulation, histogenesis, organogenesis, and eventually form something we recognize again as the adult animal. At some point there was an intrinsic preservation of the sexual event and, as Maynard Smith pointed out earlier, getting rid of these "sexual hangups" is extremely difficult. You can't drop sex any more than you could have the city of Minneapolis without electricity. I mean, you could banish electricity, but it would be extremely difficult, since so many aspects of city functioning intrinsically involve the use of electricity. Losing sex (or electricity) would require reversing all sorts of things to which commitments have already been made. In that sense, sex is intrinsic to the animal state, but it is not necessarily intrinsic to protists.

MAYNARD SMITH: It's not true that animals can't get rid of sex; it's just hard, very hard.

HAMILTON: I have some difficulty with this repair hypothesis. I can see that repair is very important, but it seems to me that it's never going to successfully explain the urge to outbreed and therefore the urge for sex. It seems to me that we began to see this in your presentation quite early on. For example, you talk of the formation of aphrodisiacs, as you call them, which causes different mating types to fuse. Can you explain why there are mating types? It doesn't seem to me that it's at all necessary for your scenario. If you're hungry, you eat anything; you don't preferentially eat just those of an opposite mating type to yourself.

MARGULIS: I completely agree that in the beginning there were no ritualized differences in mating type. Beginning with simple survival, you soon get a ritualized alternation of fusion followed by reduction. The differentiation of mating types ensures a stable alternation. I did not mean to run all the processes together, because they did not evolve together.

HAMILTON: I'm afraid I still don't see why outbreed rather than inbreed, but maybe it's far too difficult to explain here.

MAYNARD SMITH: One of the most fascinating things I learned

from your book was the existence of forms of meiosis in protists that do not involve the classic initial doubling of the chromosome number and then two meiotic divisions. It has always been a puzzle, you see, because if all you want to do is reduce the chromosome number, why start by doubling it? I wondered whether you thought that these odd forms of meiosis are ancestral to what we now see in, say, mammals, or are they a completely separate road?

MARGULIS: Heliozoans ("sun animalicules") fuse cells that are the products, of mitosis. In other words, the nuclei divide by mitosis and then fuse again. Can anything be more fundamentally inbred than that? In response to your question, Dr. Maynard Smith, heliozoans are not on the lineage to anything except heliozoans.

QUESTIONS FOR FURTHER DISCUSSION

1. Does finding cannibalism and the production of fused monsters in the present–day microcosm prove that sex arose in that fashion?
2. Is the origin of bacterial sex homologous with the origin of sex in plants and animals?
3. How does the question of the origin of sex differ from the question of why 99 percent of the animals engage in sex on this planet?
4. If by definition oddities are rare, why should evolutionists devote themselves to the study of oddities to trace the evolution of sex? Wouldn't it make more sense to explore the causes of sex in the vast majority of organisms that successfully use sexual means?

Peter Raven describes the evolutionary sexual history of plants. He approaches the question of the evolution of sex by placing sexual expression in a historical perspective. Since all plants engage in meiosis, he does not look for answers to the question of its origin in plants, but he does emphasize how sexual habits have evolved.

This address is fundamentally different from those that precede it. Here the emphasis is not on why sex, but rather on how is sex carried out. Little emphasis is placed on the possible advantages of sex until the panel discussion at the end of the address. Watch for evidence that the history of plants has constrained their sexual potentials. These constraints influence the evolutionary potential of plants and illustrate a new meaning to Maynard Smith's phrase "sexual hangups."

The Editors

3. The Meaning of Flowers: The Evolution of Sex in Plants

PETER H. RAVEN

The ancestors of today's plants first invaded the land more than 410 million years ago, and now plants are one of the dominant groups of terrestrial organisms. The origins of the plant kingdom (like the origins of all other groups of organisms) were doubtless aquatic, but plants are fully terrestrial in the expression of their adaptive patterns. A few groups of plants have secondarily invaded freshwater habitats, and even fewer (the sea grasses) are marine. The majority of the estimated 265,000 species are *vascular* plants—plants with a system of internal plumbing through which water, with its dissolved minerals and carbohydrates, moves—the remainder being bryophytes (mosses, liverworts, and hornworts). In turn, all but about fifteen thousand species of vascular plants are angiosperms, or flowering plants—by far the most important plant group today, whether their importance is measured by numbers of individuals, visual impact, or biomass.

Everywhere plants dominate and give character to the landscape; forests, scrublands, deserts, Arctic wastes—all of them are defined by the nature of their plant covers, or by the absence of plants. Except for a few tens of thousands of species of eukaryotic protists—the algae—and some bacteria, plants are the only organisms capable of photosynthesis. As such, they are fundamental to the diversity of life on Earth, as they have been since their origin. All of the 5 million (or 50 million, depending on who is doing the estimating) species of other organisms are dependent on the activities of photosynthetic organisms for the oxygen they breathe.

Plants were the first group of organisms to invade the land successfully, and they were followed later by the ancestors of the territorial arthropods (including the insects), terrestrial mollusks (there are more than 35,000 terrestrial species, or about a third of the entire group), vertebrates, and fungi—all of them dependent on plants. A number of authors have argued that the increase in oxygen in the atmosphere, with its corresponding ultraviolet–blocking ozone layer, was the critical factor that made possible the invasion of the land by plants and other organisms. By the Carboniferous Period, 360 million to 286 million years ago, the great forests that formed many of our coal deposits had evolved, and the nature of land habitats had been permanently altered.

From what group of organisms were the first plants derived? Clearly, these ancestors must be sought among the green algae, or Chlorophyta. Plants and green algae are linked by a number of unusual or even unique biochemical and structural features: for example, some green algae and all plants have cell walls dominated by cellulose; the chloroplasts of both contain chlorophylls *a* and *b*, together with characteristic carotenoids (pigments that aid in photosynthesis); and they form their reserve food substance, starch, within their chloroplasts. In addition, plant cells divide by the formation of a *cell plate*, a specialized structure that forms between dividing cells. A few groups of green algae, but no other kinds of organisms, form cell plates, which grow out to the margins of the dividing cell after forming. In all other kinds of organisms, the cells divide by pinching in from the sides. Each of these features is unusual, but taken together, they can be interpreted only in one way: plants, with their complex structure and existence on land, evolved from green algae, which have relatively simple features and are primarily aquatic.

Green algae are an extremely diverse group of at least seven thousand species. Although most green algae are aquatic, they are found in many different kinds of habitats: on the surface of snow, on tree trunks, in the soil, and as members of symbiotic relationships with fungi and some groups of protists and animals. Relatively few green algae form cell plates, and these are the only other organisms that agree with plants in other details of their cell division as well. These algae are members of the class Charophyceae. All living members of this class are too specialized in some of their features to be the actual ancestors of plants, but the specialized points of resemblance between the two groups are so extensive that we can be sure that the ancestor of plants would be regarded as a member of the Charophyceae if it were known.[1]

In order to compare plants with their ancestors in more detail, and to discuss the evolution of their sexuality, it is necessary first to discuss the nature of life cycles in general. In animals, adult individuals are *diploid*, containing twice the number of chromosomes in the eggs and sperm. In human beings, for example, most cells contain forty–six chromosomes, whereas the eggs and sperm contain twenty–three chromosomes each. Such a reduced number of chromosomes is said to be *haploid*. The gametes–*gamete* is the collective term for eggs and sperm—are formed as a result of the process of meiosis, and they fuse to form a cell called the *zygote*, in which the diploid chromosome number is again restored. Once it is formed, the zygote divides by a process called *mitosis*—a process that does not involve a change in chromosome number, and which is characteristic of all living cells except for those of the bacteria. If the particular kind of organism is multicellular, as in animals or plants, then the zygote divides by means of mitosis to give rise to the body of the adult, which often consists of millions of cells, some of which may be very different from one another in their features.

The most fundamental difference between the life cycle of an animal and that of a plant is that in an animal the gametes are the only haploid cells present in the life cycle. Once they are formed, they fuse, forming a zygote, and the zygote divides to produce the body of the individual animal. In plants, on the other hand, the haploid reproductive cells formed by the diploid individual are not gametes; rather, they are *spores*. Instead of fusing, these spores divide by *mitosis*, producing haploid individuals. These haploid individuals, called *gametophytes*, in turn eventually produce gametes, which function like those of animals, fusing and giving rise to a zygote, which is the first cell of the *sporophyte* generation.

Summarizing these differences in another way, an animal is diploid; it produces gametes, following meiosis; and these gametes fuse, producing the zygote, the cell that ultimately develops by mitosis into the adult animal. In vascular plants, the individuals that you usually see—trees, shrubs, grasses, herbs, ferns—are sporophytes—diploid individuals—comparable in this respect to adult animals. When meiosis, or reduction division, occurs in specialized organs on these individuals (and we will discuss the details later) it leads to the production of spores—haploid cells that divide by mitosis, giving rise to a *haploid*, multicellular individual called a gametophyte, from which the gametes ultimately develop. In animals, the progression is adult → meiosis → gametes → syngamy (cellular fusion) → adult; in plants, the progression is "adult" sporophyte → meiosis → spores → gametophyte → gametes → syngamy → zygote → adult (see fig. 1).

With this background, we can now consider the sexual features of Charophyceae in relation to those of the plants. First of all, like both animals and plants, the Charophyceae are *oogamous:* their female gametes are nonmotile, and therefore called eggs; whereas their male gametes are motile, swimming by means of flagella, and are called sperm. In many green algae, both female and male gametes are motile, and in most of these, they are equal in size and similar in appearance. Oogamy has originated several times within the group, and was certainly characteristic of the ancestors of the plants.

Second, in the Charophyceae that gave rise to the plants, the egg was retained on its parent until fertilization took place, rather than being shed into the water. This feature still occurs in some Charophyceae and is universal in plants; it therefore was almost certain to have been characteristic of the ancestor of the group.

Third, in some species of *Coleochaete*, a genus of Charophyceae that has many plant–like features, the sperm are formed within a multicellular, internal structure. Such a structure, known as an *antheridium*, is found in primitive plants, but has disappeared many times during the course of plant evolution, and is missing in most living plants.

Fourth, the zygote, (the only diploid cell that is formed during the life cycle of the organism) is held on the parent individual in some species of *Coleochaete* as it is in all plants. Such retention ultimately led to the evolution of the *embryo*, an important feature in the life cycles of all plants, and one that we shall subsequently discuss in detail.

As Linda Graham[2] has pointed out, perhaps the most distinctive feature of the land–plant reproductive cycle is that eggs are retained on the parental gametophyte and fertilized there. The resulting embryo, or young sporophyte, remains associated with its parent gametophyte and derives nutrition from it during early

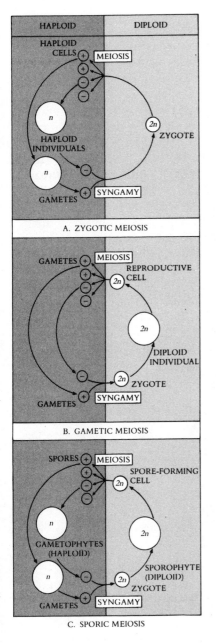

Figure 1. Diagrams of three major kinds of life cycles in eukaryotes. Top to bottom they are (A) *zygotic meiosis,* a kind of life cycle in which the zygote is the only diploid cell; found in fungi and many unicellular organisms; (B) *gametic meiosis,* in which meiosis produces the gametes, which fuse, giving rise to a zygote; found in most animals, the gametes are the only haploid cells; (C) *sporic meiosis,* in which there is a regular alternation between a multicellular haploid phase and a multicellular diploid phase. The diploid phase produces spores that give rise to the haploid phase, and the haploid phase produces gametes that fuse to form the zygote. The zygote is the first cell of the multicellular diploid phase. This kind of life cycle is characterized by *alternation of generations,* and is characteristic of plants. (Reprinted by permission from *Understanding Biology* by P. H. Raven and G. B. Johnson, [St. Louis MO: C. V. Mosby and Co.], figure 24–14, page 479.)

development (see fig. 2). This complex feature is so important, in fact, that plants are often called *embryophytes*—embryo–producing plants. In the seed plants (gymnosperms and angiosperms among living groups), the gametophyte in turn is held within a drought–resistant structure, the seed coat, and the embryo develops within a seed.

In the algae, on the other hand, eggs and zygotes are rarely retained on their parents; they are usually released into the water, developing there independently. There is no nutritional or developmental dependence between the haploid and diploid components of the life cycles of green algae, and consequently no embryos. How did this important plant feature arise?

In *Coleochaete*, as we have seen, the diploid zygotes are retained on their haploid parents, a feature that is very unusual among green algae. A classical problem of plant evolution has been whether their ancestors already had an alternation of generations, as proposed more than a century ago by the Czech botanist L. Celakovsky; or whether the sporophyte generation arose from the elaboration by mitosis of a retained zygote like that of *Coleochaete*. The latter theory was first proposed by the great English

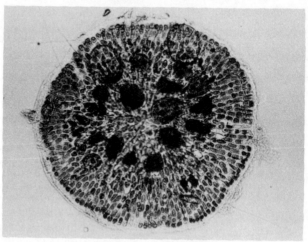

Figure 2. *Coleochaete soluta*, freshwater alga that shares many features with plants. Large, dark spots near the center of the disc are zygotes, which is one of the features of *Coleochaete* that is especially plant–like. (Courtesy of Linda E. Graham, University of Wisconsin, Madison.)

plant morphologist F. O. Bower in 1908. Some green algae do indeed have an alternation of generations, but in them, the gametophytes and sporophytes are free–living and not dependent on one another. The dilemma presented by accepting them as the ancestors of plants, therefore, concerns the way in which the nutritional and structural dependence between the generations that is shared by all plants arose.

Fossils that have been discovered and studied over the past half–century have revealed a considerable antiquity for green algae that resembled *Coleochaete,* and have thus strengthened Bower's hypothesis. In addition, the structural details that have been mentioned above make a close relationship between Charophyceae such as *Coleochaete* and plants almost certain, a result that is also in line with the acceptance of this theory of plant origins. No living green algae that possess an alternation of generations have the plant–like characteristics of the Charophyceae; none, therefore, seems likely to have been the ancestor of the plants. Considering that no living Charophyceae has an alternation of generations, it seems virtually certain that the sporophyte generation of plants did arise by the elaboration of the retained zygote of ancestral Charophyceae more than 410 million years ago.

In the evolution of alternation of generations in plants, as Linda Graham[3] has pointed out, there would have been four steps.

1. The evolution of differences between the gametes, in which one, the egg, became larger and immobile while the other, the sperm, became smaller and mobile.
2. The retention of the egg on the parent gametophyte at the time of fertilization
3. The retention of the zygote and its subsequent development on the gametophyte.
4. The establishment of a nutritional and developmental relationship between sporophyte and gametophyte.

As Dr. Graham visualized it, the evolution of dependent, multicellular sporophytes may have occurred in an aquatic habitat. Most species of *Coleochaete* occur in shallow water, where they grow on the stems of submerged plants, on rocks, or on other submerged objects. In such circumstances, the retention and development of the sporophyte on its parental gametophyte would tend to keep the organism in a favorable habitat, near suitable substrates for the establishment of the spores of the next generation. The tightly associated complex of gametophyte and sporophyte cells that is characteristic of some species of *Coleochaete* seems likely to provide a model for the kind of situation that could have given rise to alternation of generations as found in plants. The relationship found in plants is so complex, both developmentally and nutritionally, that it seems highly unlikely to have originated more than once.

The first plants, therefore, seem to have formed embryos that were held within the tissues of their parent gametophytes and derived nutrition from them. As in contemporary bryophytes, ferns, and a few other groups, the sperm probably developed within multicellular structures, antheridia, and then were released, swimming through water to the egg. As pointed out by Bruce Tiffney,[4] oogamy (the specialization of the gametes into eggs and sperm) is favored in a drying environment, such as that on land, because only half of the gametes (the sperm) are subjected to drying out. The retention of the young sporophyte by the gametophyte is clearly favored for similar reasons, and the stage is set for the evolution of simpler modes of dispersal. This in turn sets the stage for the mode of fertilization that evolved among plants during their course of extraordinary evolutionary radiation on land.

All mosses, liverworts, and hornworts have sporophytes that are somewhat nutritionally dependent on the gametophyte. Both in hornworts and in most mosses, the sporophytes possess chlorophyll and are photosynthetically active, however; they do not normally grow independently of the gametophytes. In ferns (a vascular plant) the gametophyte is a delicate, usually flattened plant, often about a centimeter long, that grows in protected damp places. In contrast to the mosses, liverworts, and hornworts, the sporophytes of ferns start to grow within the tissues of the gametophyte and ultimately become both nutritionally and developmentally independent. It is the sporophyte that develops into the fern plants that are familiar in our woods and gardens.

Among a few kinds of ferns, the sporophytes produce two kinds of spores: large ones that develop into egg–producing gametophytes, and small ones that develop into sperm–producing gametophytes. This evolutionary development is perpetuated in all of the seed–forming plants, or seed plants, the gynmnosperms and angiosperms (flowering plants). In them, the gametophytes are reduced in size and are contained within the tissue of the sporophytes and nutritionally dependent on them. Mature gametophytes in most seed plants are less than a millimeter in length. They have been progressively reduced during the course of evolution so that the mature sperm–producing gametophyte of a flowering plant is a germinated pollen grain, with only three cells, and the mature egg–producing gametophyte usually has only eight nuclei and is contained within the seed–producing structures of the plant. Nonetheless, these highly reduced structures are homologous with an entire individual of *Coleochaete*, excluding the zygote, which is the only diploid cell in the life cycle of the Charophyceae, or homologous with the green, visible moss plants that you may see on rocks or in the woods.

Sex among the liverworts, mosses, hornworts, ferns, and several other groups of vascular plants is rather straightforward. These plants produce their eggs by mitosis and then retain them in bottle–shaped structures called *archegonia* that are borne on the gametophyte. The sperm, produced in antheridia, are released and

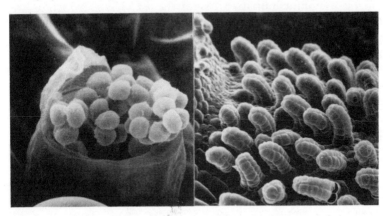

Figure 3. Sexual organs on gametophytes of the fern *Anemia mexicana*. At the left, a sporangium is shown releasing sperm, which are the only flagellated cells in the life cycle; at the right are shown a number of flask––shaped archegonia. They will open at the apex at maturity; each contains a single egg. (Courtesy of Joan E. Nester.)

swim through free water to the mouth of an archegonium, enter it, and swim down within, ultimately fusing with the egg (see fig. 3). The young sporophyte divides by mitosis and, as we have seen, ultimately becomes nutritionally independent of the gametophyte in ferns and other relatively simple vascular plants. The psilophytes, clubmosses, and horsetails all exhibit this kind of life cycle, with both phases nutritionally independent, although some extinct clubmosses were seed plants and the nutritional status of their gametophytes apparently varied.

The situation in the gymnosperms and angiosperms, the living seed plants, is different. In some gymnosperms, the massive, cellular female gametophyte produces archegonia; in others, it simply produces eggs directly within its tissues. In either case, the female gametophytes are held within the tissue of the sporophyte in structures called ovules. No angiosperms have archegonia; their female gametophytes are highly reduced; a single egg is produced within each female gametophyte. The phenomenon of *double fertilization* is characteristic of all angiosperms: one sperm nucleus fertilizes the egg, producing the zygote, or first cell of the sporophyte generation, while the other sperm nucleus fertilizes typically two of the other nuclei of the female gametophyte, producing a unique nutritive tissue, the *endosperm*, which is triploid—each cell contains three sets of chromosomes. In both gymnosperms and angiosperms, however, the ovules mature into seeds after fertilization. In angiosperms, seeds are contained in fruits, which we shall discuss later; in some gymnosperms, they are borne on the surface of scales in a cone, while in others, they are simply naked.

The mature male gametophytes of both gymnosperms and angiosperms, as we have mentioned, are germinated pollen grains. Pines have six cells or nuclei in the mature male gametophyte, while angiosperms have only three, two of them taking part in the process of double fertilization. The pollen of seed plants is shed from the parent sporophyte before it is mature, and becomes fully mature only after germination. Pollination is the transfer of pollen from the organs that produced it, which may be on the same or a different plant, to the vicinity of the egg; it is therefore *not* synonymous with fertilization. One curious fact is that in the cycads (gymnosperms with fernlike leaves) and in ginkgo (the maidenhair tree) the sperm are still multiflagellated and therefore mobile; in other living gymnosperms and in the angiosperms, they are not. At any rate, they are carried passively to the vicinity of the egg in both kinds of seed plants, protected from desiccation within their tough, highly drought–resistant pollen wall. The ways in which pollen is transferred and the rules that govern its production dominate the subject of sex among seed plants, a group of organisms that has been dominant for well over 200 million years.

In most gymnosperms, pollen is carried passively in the wind from the male cones (those which produce pollen, and therefore, male gametophytes) to the ovules (structures that contain the female gametophytes). In recent years, Karl Niklas of Cornell University[5] has produced a series of illuminating studies of the aerodynamics of wind pollination, and demonstrated convincingly that many wind–pollinated plants actually manipulate the wind, their structures having evolved to improve their chances of successful pollination. For example, a pine cone acts like a turbine, causing pollen–containing air to spiral around it; at the same time, air currents between the scales carry pollen grains into the vicinity of the ovules. The spirally arranged scales of the cone, in effect, cause the cone to wrap itself in a blanket of moving air. Other plants, including wind–pollinated angiosperms, produce channels along which air is directed to the vicinity of the ovules or *stigmas*, the pollen–trapping structures in angiosperms. The morphological features, especially density and size, of the tiny pollen grains also affect their aerodynamic properties and the frequency with which they are carried to the ovules of particular plants. Niklas has also analyzed wind–pollinated angiosperms, such as grasses, from a similar point of view, clearly demonstrating the importance of such considerations.

In order to understand the role of pollination in the evolution of angiosperms, it is necessary first to consider the probable antecedents of the angiosperms. Extinct groups of gymnosperms known as Bennittitales, *Caytonia,* and the corystosperms appear most closely related, while modern evidence has shown that, among living seed plants, a small group known as Gnetales certainly includes the closest relatives of angiosperms.[6] All seed plants seem to have been derived from ancestors with fern–like leaves, and the precise ancestors of angiosperms are not yet known. At any rate, double fertilization is a unique feature shared by all angiosperms. Moreover, in all members of this group, the ovules are enclosed within a *carpel*, which is completely closed in all but a very few members; and most angiosperms have a specialized *stigma*, with receptive tissue to which pollen grains adhere, and on which they may germinate. Stigmatic fertilization, a closed carpel, and double fertilization are therefore features indicating that all angiosperms had a common origin.

What about interactions with animals? Chaloner[7] pointed out that the very first land plants, more than 400 million years old, are associated with arthropods, which were probably both eating their spores and vegetative parts and seeking shelter in their stems and reproductive structures even at that remote time. Spore–feeding (and pollen–feeding) therefore seems to have preceded the evolution of the seed and to have provided a background against which its evolution took place. Angiosperms almost certainly originated before their first documented appearance in the fossil record, but when they did so remains unknown. Judged from their structures and from evidence gained from studying living gymnosperms, insect pollination certainly originated before the angiosperms appeared. The particular features of the angiosperm flower, which define the group, and especially the enclosed ovules, apparently made the early angiosperms more efficient in this respect.

In the flowers of primitive angiosperms, ovule–producing structures—carpels—and the pollen–producing sturctures—stamens—are combined. In contrast, wind–pollinated gymnosperms and angiosperms, in which this system of pollination seems to have evolved secondarily, characteristically have separate ovule–producing and pollen–producing structures, sometimes even borne on separate plants. As Chaloner and others have pointed out, pollen–feeding insects would bring about pollination only if the ovule–producing structures were located nearby. A number of authors have postulated that the enclosure of the ovules would have been advantageous to these early insect–pollinated plants, since the insects might otherwise devour the ovules and thus lower the reproductive capacity of the plants. This, and possibly other factors, would have combined to give the ancestral angiosperms a competitive advantage and brought them to world dominance approximately 100 million years ago. Somewhat surprisingly perhaps, many angiosperm groups, including grasses and Juglandaceae, reverted to wind pollination before the close of the Cretaceous Period. This indicated that at least in circumstances where large, relatively homogeneous populations of a plant can grow, wind pollination is advantageous.

In the bisexual flowers of angiosperms, a number of devices that led to enhanced outcrossing evolved early in the history of the group. Among these features was genetic self–incompatibility, by virtue of which an individual is incapable of pollinating itself successfully. The pollen tubes may be inhibited on the stigma, or in the tissue of the carpel; fertilization may fail to take place; or the embryos may not develop properly. Alternatively, the anthers and stigmas may be spatially separated in the flowers, thus decreasing the chance of pollen reaching the stigmas within a flower and increasing its chance of being carried to another flower and perhaps another plant. Such systems, which are called *herkogamous* ones, also restrict the frequency of useless self–pollination in genetically self–incompatible plants.[8] In *dichogamous* plants, the pollen is produced either before or after the period when the stigmas are receptive.[9] to mention a single extreme example, when an avocado flower first opens, the stigma is receptive and the anthers have not yet split to shed their pollen. After remaining open for several hours the flower closes, but it opens again the next day. At that time, the stigma is dried up and the anthers are shedding pollen: the flower is functionally staminate.

In a number of angiosperms, the flowers contain *only* functional stamens *or* functional carpels, but not both. Such flowers are said to be staminate or pistillate, respectively. They may occur together on an individual plant; in such cases, the plant is said to be *monoecious*, from the Greek words meaning "one household." Corn and oaks are examples of monoecious plants; but oaks are also largely genetically self–incompatible, thus repressing self–pollination even if the pollen produced by an individual tree does reach its pistillate (acorn–producing) flowers. In other plant species, such as willows, the pistillate and staminate flowers are born on different individuals; such plants are said to be *dioecious:* they have "two households."

Genetic self–incompatibility seems to have evolved among the very earliest angiosperms or their ancestors, judged from its distribution among living plants.[10] There are no known examples of genetic self–incompatibility in gymnosperms. Much more research is needed in this area, and information has accumulated

Figure 4. *Eupomatia* represents a primitive family of angiosperms found in northeastern Australia and eastern New Guinea. Its large, multiparted flowers are pollinated primarily by weevils, a family of beetles of which one is shown here at the flowers. (Courtesy of Trevor Hawkeswood.)

Figure 5. *Zygogynum baillonii* is a member of a very primitive family of angiosperms. The reddish–brown flowers of this genus, which is found only on the island of New Caledonia in the south Pacific Ocean east of Australia, are pollinated by primitive, pollen–eating moths of the genus *Sabatinca*, shown at the right. (Courtesy of Leonard Thien.)

slowly, primarily because of the tropical and subtropical distribution of many living primitive angiosperms. Whether all systems involving genetic self–incompatibility in angiosperms had a common origin is doubtful,[11] but can be tested—especially with the availability of modern molecular approaches to the subject.[12] Zavada and Taylor[13] have hypothesized that the selective impetus for the evolutionary development of the closed carpel, one of the leading features of angiosperms, was the origin of genetic self–incompatibility. In their view, genetic self–incompatibility provides a means for the female to evaluate the quality of male gametes as the pollen tubes germinate and grow through the stylar tissue. This is an alternative to the traditional view, mentioned above, that the ovules were enclosed to protect them from the chewing mouthparts of early pollinators.

What were the characteristics of the flowers of the first angiosperms, and how were they pollinated? There are at least three groups of pollination systems among existing primitive angiosperms, as outlined by Leonard Thien:

1. *Plants with large, multiparted flowers:* In these primitive angiosperms the pollen is often eaten by beetles (see fig. 4) and flies, exceptionally by moths (see fig. 5), and these insects may mate in the flowers as well.[14] By position, as well as by movements during the maturation of the flower, the stamens and stigmas are often separated. In addition, they are often provided with petal–like structures (staminodes) intermediate between petals and stamens, which may be scented or colored so as to attract insects. Included here are magnolias, custard apples, and most water lilies, among others.

2. *Plants with small, simple, open flowers:* These plants use fly and beetle pollination as judged by their floral structure.[15] Wind pollination may also occur in some groups. Included here are the families Amborellaceae, Chloranthaceae, and Trimeniaceae.

3. *Plants with small, unisexual, somewhat closed flowers:* These plants are presently visited by beetles and flies. Included here are the nutmeg family (Myristicaceae) and Monimiaceae.[16]

As this list demonstrates, beetle and fly pollination predominate, with wind pollination present in a few genera of primitive angiosperms. Both beetles and flies (generalized ones, not the specialized flower–visiting members of either group that are now frequent at flowers) greatly antedate the Cretaceous angiosperms, and were probably associated both with the ancestors of angiosperms and with the earliest members of the group. The flowers of most primitive angiosperms are not brightly colored, and they are generally radially symmetrical. Floral odors seem to have evolved very early, judged from their prevalence among primitive angiosperms; they were probably the first important attractant for pollinating insects. The fact that the mating of insects occurs on the flowers of many primitive angiosperms probably indicates an additional means whereby the specificity of visits to individual kinds of flowers could have been established.

From these primitive angiosperms evolved flowers with many different kinds of distinctive characteristics. Pollination by bees, for example, now so well represented among the flowering plants, obviously, evolved subsequent to the late Cretaceous appearance of the bees. As Bernhardt and Thien[17] have pointed out, pollen that is usually retained within the anthers following their dehiscence and the presence of narrow, often thread–like filaments in the anthers, in contrast to the broad and somewhat leaf-like anthers found in many primitive angiosperms, appear to be features that evolved in connection with pollination by bees. More and more angiosperm flowers are being discovered in the fossil record, even from the Cretaceous Period[18]; and as their features become better known, we shall learn a great deal more about when the different kinds of features of flowers evolved. We know, for example, that nectaries existed in some of the flowers of the Upper Cretaceous, a feature that suggests the presence of specialized pollination systems.

In summary, the evolution of bilaterally symmetrical and highly irregular flowers, common among contemporary angiosperms, clearly took place mainly during the past 50 million years. As such flowers evolved, so did the complex and fascinating interactions between them and their pollinators. Angiosperms, being rooted to the ground and thus immobile, control their choice of mates by their morphological and chemical features and the ways in which they present them. Their remarkable success in doing so has played a major role not only in their adaptive radiation into the approximately 250,000 species that dominate land habitats today, but also in the comparable adaptive radiation of the animals involved in these interactions. The systems involve an array of complex and interesting systems which are the subject of intensive study by hundreds of scientists around the world. As the theoretical basis of such studies is improved, they will be carried out with increasing precision, ultimately revealing much more than is even suspected now about the evolution of a major segment of life on Earth.[19]

NOTES

1. Mishler, B. D., and S. P. Churchill. 1985. Transition to a land flora: Phylogenetic relationships of the green algae and bryophytes. *Cladistics* 1:305–328; Sluiman, H. J. 1985. A cladistic evaluation of the lower and higher green plants (*Viridiplantae*). *Plant Systematics and Evolution* 149:217–232; Bremer, K. 1985. Summary of green plant phylogeny and classification. *Cladistics* 1:369–385.

2. Graham, L. E. 1985. The origin of the life cycle of land plants. *American Scientist* 73:178–186.

3. Ibid.

4. Tiffney, B. H. 1981. Diversity and major events in the evolution of land plants. In: *Paleobotany, Paleoecology and Evolution.* K. J. Niklas, ed. Volume 2. Praeger Publishers: New York. 193–230.

5. Niklas, K. J. 1985. The aerodynamics of wind pollination. *Botanical Review* 51:328–386; Niklas, K. J. 1985. Wind pollination—A study in controlled chaos. *American Scientist* 73:462–470.

6. See, for example, Doyle, J. A., and M. J. Donoghue. 1987. The origin of angiosperms: A cladistic approach. In: *The Origin of Angiosperms and Biological Consequences.* E. M. Friis, W. G. Chaloner, and P. R. Crane, eds. Cambridge University Press: Cambridge, England; Doyle, J. A., and M. J. Donoghue. 1987. The importance of fossils in elucidating seed plant phylogeny and macroevolution. *Review of Paleobotany and Palynology* 50:63–95.

7. Chaloner, W G. 1984. Plants, animals and time. *The Paleobotanist* 32: 197–202.

8. Webb, C. J., and D. G. Lloyd. 1986. The avoidance of interference between the presentation of pollen and stigmas in angiosperms. II. Herkogamy. *New Zealand Journal of Botany* 24:163–178.

9. Lloyd, D. G., and C. J. Webb. 1986. The avoidance of interference between the presentation of pollen and stigmas in angiosperms. I. Dichogamy. *New Zealand Journal of Botany* 24:135–162.

10. Bernhart, P., and L. B. Thien. In press. Self–isolation and insect pollination in the primitive angiosperms: New evaluations of older hypotheses. *Plant Systematics and Evolution;* Zavada, M. S., and T. N. Taylor. 1986. The role of self–incompatibility and sexual selection in the gymnosperm–angiosperm transition: A hypothesis. *The American Naturalist* 128: 538–550.

11. Gibbs, P. E. In press. Do homomorphic and heteromorphic self–incompatibility systems have the same sporophytic mechanism? *Plant Systematics and Evolution.*

12. Anderson, M. A. *et al.* [17 authors]. 1986. Cloning of cDNA for a stylar glycoprotein associated with expression of self–incompatibility *in Nicotiana alata. Nature* 321:38–44.
13. Zavada, M. S., and T. N. Taylor, 1986. The role of self–incompatibility and sexual selection in the gymnosperm–angiosperm transition: A hypothesis. *The American Naturalist* 128:538–550.
14. Gottsberger, G. 1977. Some aspects of beetle pollination in the evolution of flowering plants. *Plant Systematics and Evolution*, Supplement 1:211–226; Pellmyr, O., and L. B. Thien. 1986. Insect reproduction and floral fragrances: Keys to the evolution of the angiosperms? *Taxon* 35:76–85; Thien, L. B., P. Bernhardt, G. W. Gibbs, O. Pellmyr, G. Bergstroem, I. Groth, and G. McPherson. 1985. The pollination of *Zygogynum* (Winteraceae) by a moth, *Sabatinca* (Micropterygidae): An ancient association? *Science* 227:540–543.
15. Endress, P. K. 1986. Reproductive structures and phylogenetic significance of extant primitive angiosperms. *Plant Systematics and Evolution* 152:1–28.
16. Endress, P. K. 1980. Ontogeny, function and evolution of extreme floral construction in Monimiaceae. *Plant Systematics and Evolution* 134:79–120.
17. Bernhart, P., and L. B. Thien. In press. Self–isolation and insect pollination in the primitive angiosperms: New evaluations of older hypotheses. *Plant Systematics and Evolution.*
18. See, for example, Friis, E. M. 1982. *Scandianthus* gen. nov., angiosperm flowers of saxifragalean affinity from the Upper Cretaceous of southern Sweden. *Annals of Botany* 50:569–583; Friis, E. M. 1984. Preliminary report of Upper Cretaceous angiosperm reproductive organs from Sweden and their level of organization. *Annals of the Missouri Botanical Garden* 71:403–418.
19. I acknowledge the very kind assistance of Linda Graham of the University of Wisconsin, Madison; Leonard B. Thien, Tulane University; Peter K. Endress, Institut für Systematische Botanik der Universität Zürich, who provided extremely valuable information, views, and literature reviews in the preparation of this paper.

PANEL DISCUSSION

MARGULIS: What is the selection pressure, if any, to keep plants outbreeding?

RAVEN: Plants are enormously flexible in their outcrossing and inbreeding and in their breeding systems in general. They are much more flexible than vertebrates. Plants have the whole panorama available to them, from 100 percent obligate outcrossing, to 0 percent outcrossing. Outcrossing species tend to grow in broadly defined favorable habitats where the plants are not exposed to enormous stresses or where the habitats are heterogeneous and outcrossing is important to produce variable offspring. While that is the ecological pattern, I find it very difficult to apply that to a question of why a particular species engages in 100 percent outcrossing.

MAYNARD SMITH: One thing that plants and animals do have in common is that inbreeding is extremely bad for them. If you inbreed a *Drosophila* population its fitness will go down by a factor of one hundred in about five generations of brother–sister mating, which is equivalent to about three generations of selfing. I don't see any difficulty in explaining why plants outcross. They outcross to avoid inbreeding depression, no question. It is a more complex question to ask why some plants self. If you consider our inbreeding *Drosophila* population again, after the first five generations of inbreeding it will begin to recover. It will recover because it gets rid of all of its deleterious recessives. The inbreeders that we see around us are those that, for one reason or another, have gotten through the threshold of inbreeding depression and are struggling up the other side.

RAVEN: Plants do seem to have gotten through the threshold of inbreeding depression (or through a bottleneck) quite frequently. When they do, it is very advantageous for the survivors. We see genera of plants that are very widespread but also very inbred. This is just another way of making the point that plants are very versatile.

HAMILTON: I don't feel that inbreeding depression can be a sufficiently universal explanation to account for the emphasis on outbreeding, because there are haploid plants that consistently outbreed. For example, the mosses that Dr. Raven described have incompatibility mechanisms that prevent them from inbreeding.

Now it is true that they do produce the diploid sporophyte, but the sporophyte is a comparatively small, dependent organism which one would have thought could have gotten away with quite a bit of genetic failure, and yet they still insist on outbreeding. Moving down a little lower, of course, are the yeasts, which often spend most of their time in the haploid state, yet also have mating types that prevent them from inbreeding. So my feeling is that some other mechanism must be invoked in explanation. I don't deny the existence of inbreeding depression as being a short-term factor, but I think some more universal cause needs to be sought.

AUDIENCE QUESTION: The Red Queen hypothesis suggests that long–lived organisms must "run" harder than short–lived species. Do long–lived plants have special ways to speed recombination?

RAVEN: Long–lived plants are much more highly outcrossing than short–lived ones. There is a perfect correlation between longevity and outcrossing in plants. Self–pollinating plants are much better represented among annuals.

Graham Bell tells me that chiasma frequency (a measure of crossing over frequency during meiosis) is highest in long–lived trees and habitual selfers. The point is that habitual selfers, while they may be quite short–lived, have very few opportunities for recombination. When they self, recombination does nothing, so in effect they are plants with a very long generation time.

QUESTIONS FOR FURTHER DISCUSSION

1. In what ways are the sexual habits of flowering plants the outcome of their history rather than being the best solution to present–day challenges?
2. Some researchers have suggested that frequent somatic mutations are more common in long–lived plants than in annuals. Is this observation consistent with Raven's answer to the Red Queen question?
3. Can the self–incompatibility of mosses be a kind of "sexual hangup," or does the history of the lineage suggest that it is a recent modification of the sexual practices of these primitive plants?

In this chapter, Hamilton develops the idea that sex compensates for the inherent disadvantages large, long–lived organisms face when locked in a "coevolutionary arms race" with more rapidly evolving disease organisms. The structure of his argument is fundamentally different from that of Margulis and Sagan, who stress the history of sexual origins and emphasize "sexual handups" as the basis for sex as we know it today. Hamilton takes a different approach arguing that the expression of sex in present–day populations is advantageous in today's environment.

Hamilton is a skilled evolutionary biologist whose mathematical models have provided insight into the dynamics of evolutionary change in populations. He is most widely known for his idea of "inclusive fitness," a guiding concept for understanding evolution in social situations. The facility with which he blends conjecture and evidence allows him to move beyond simple descriptions of natural history to see the broad outlines of the evolutionary process.

Although Hamilton points out that diesease is the only single factor ubiquitous enough to account for sex, critics have suggested that his arguments are not, in fact, directly applicable to all sexual species. Are all the sexual members of the microcosm plagued by diseases? Do plants that are free from disease violate Hamilton's hypothesis? As you read Hamilton's address, evaluate for yourself whether these criticisms are well founded. Hamilton is such a persuasive writer and prominent evolutionist that it is difficult not to agree with him, but history should not bias your judgment.

At the end of his address, Hamilton takes a refreshingly nonconformist view of what lies ahead with regard to sex. His "quaint faith" in sex forces one to reevaluate his whole argument in light of the pressing problems of human social welfare and human health. This section is not just window dressing meant to keep the lay audience attentive. Be aware of the subtle issues being raised as to the function of sex.

The Editors

4. Sex and Disease

WILLIAM DONALD HAMILTON

Nothing 'gainst time's scythe can make defence
Save breed, to brave him when he takes thee hence.
SHAKESPEARE, Sonnet XII

The decade that is witnessing the emergence of a new lethal, sexually transmitted disease may seem a bad one in which to be supporting a theory that sex is life's main adaptation to combat disease. But such a theory exists, and it seems to me the best contender in a great puzzle.[1] The theory began before AIDS,[2] and my contribution to this symposium will be devoted to showing the arguments and the evidence.

"Can you be claiming we need sex to create immunity to diseases that are spread by having sex—this must be nonsense!" This seems a reasonable enough reaction at first, and certainly the venereal diseases are going to have be reckoned on the debit side of any such general adaptation, as I will claim sex to be. But of course the number of human diseases that are transmitted sexually is actually quite low compared to those transmitted in other ways. It just happens that today most of the other deadly diseases, such as plague, smallpox, and cholera, have been on the wane for some time due to progress in medicine and public hygiene, so the new one which is venereal looms very large. However, many of the others are very far from negligible even now. Malaria has long been one of humankind's greatest natural killers, apart from the diseases of old age; it still is and indeed is resurgent. Outside of Europe and America, certain enteric and pneumonic diseases are still very serious, especially as infant killers; and then there are measles, hepatitis, dengue, all the parasites that may not kill but certainly weaken, and so on.[3] Looking back merely two centuries in Europe, a theory that took for granted that disease was one of the most common, pervasive, and terrible of the threats to human life would have been easy to accept. Two hundred years ago, a theory that claimed in essence to explain why the hand of death seemingly so arbitrarily plucks some members of a family while sparing others equally exposed, and why the hand of death plucks some families entirely while it spares others equally entirely, and why it favored some races against others—this theory might then at least have had the benefit of appealing to common experience. In old literature, one does not have to be reading Defoe's account of the great plague in London in order to see this: abundant reference to the ever–present threat can be found in the writing of almost any author from Chaucer to Dickens.

I took up what is really the very peripheral issue of AIDS in humans in order to emphasize the seriousness of disease in general; but let us just consider for a moment whether the issue of genetic resistance is really so passé, even now and even for AIDS. May we not have a prospect of a common experience, just as in the Middle Ages? Please, for a moment, take the imaginative step of allowing my idea—that is, assume that sexual reproduction might serve to bring together new combinations of genes for offspring that could resist current important *nonsexual* diseases. Surely it is plausible that such a system will also work to alleviate its own inevitable hangers on: the sexually transmitted diseases. Such a possibility seems actually quite strong in the case of AIDS.

Clinical genetic studies, published in the last few months,[4] indicate that you may indeed be able to reduce the chance that your offspring will die of AIDS if you select your partner on the basis of certain well–defined, biochemically identifiable, genetic traits. You yourself may already have the lucky genotype; or you may have one of the less lucky ones; or you may have the very bad one. It seems, if this report is right, that there may well be today some 8 percent of Europeans who would never die of AIDS, or even develop any signs of infection after being constantly exposed to it.

Can it be just by chance that there existed in the human population a common set of neutral genetical variants for a protein— that is, a neutral polymorphism—that profoundly affects susceptibility to the AIDS virus, and that this polymorphism was drifting under no selection right up to the time when the virus makes its first appearance—that is, drifting for millions of years? This seems incredible. The particular immunoglobulins that are coded by the genes that I am talking about were already under investigation before their associations with AIDS were noticed. This was because of intrinsic interest—their involvement with vitamin D transport and certain aspects of cell communication in the immune system—and also because of some other weak associations with obscure diseases like rheumatoid arthritis. Thus it seems to me far more likely that they were already involved with disease defense, most likely with defense against other viruses, perhaps ones that are not currently very pressing or dangerous. This is only a guess, I have to admit, but it can serve to lead to my next point.

It is a necessary part of the theory of sex that I am presenting that there is no feasible complete genetical answer to the disease threats pressing on a species, nor even any moderately satisfactory answer that remains good for any length of time. Disease threats are in process of continual change. Thus if AIDS were to remain medically uncontrolled (and people were to remain foolish about the obvious behavioral measures), the end of the epidemic might leave the frequencies of the genes just mentioned at completely different values from what they have today—the immune genotype mentioned above, for example, might go from its present 8 percent to nearer, say, 80 percent. But maybe a decade or two later the genotype that was the *worst* for AIDS might prove the *best* when a new virus spreads to human beings. The frequencies might swing back, or perhaps they would swing towards some quite new state.

This would be natural selection, but it is nothing like an answer to the problem of sex. A human population that reproduced by parthenogenesis could respond in very much the same way and would probably achieve a given adjustment of frequencies faster.[5] For a theory of sex based on disease resistance to have any hope at all, it must discuss not just a single locus but a minimum of several, and the alleles at the different loci must be interacting in some interesting way. It has to be the advantage of special *combinations* of alleles that favors having sex. New gene combinations are sex's specialty; indeed in nuclear genetic terms, new combinations are perhaps all that sex can supply, although outside of the nuclear chromosomes there are other interesting cytogene possibilities that still deserve study.[6]

Take the viewpoint of a gene in a very common situation, housed in a chromosome of a female in a species where the male does little or no work for the brood. Imagine this is also a species where there is an option of parthenogenesis—the female doesn't have to reproduce sexually, but she can. Species in which such an option exists are actually extremely rare, and there are fairly easily seen reasons why populations will quickly move off such an evolutionary knife edge if they even come to alight on one;[7] but let us assume that the option exists. Suppose you are the gene that has the option—you work the switch that determines whether your carrier is going to be parthenogenetic or not. What sort of information would you want to gather to decide which was the best option (that is, either sex or perfect replication of your bearer's genotype) giving regard to the long term survival of copies of yourself?

If things seemed to have gone well during the lifetime of the population of cells that is your present bearer, and if your information about the environment suggested it was going to stay the same, then I think you would undoubtedly choose parthenogenesis. After all, if you get mated, there is a 50 percent chance for any particular offspring that you won't be transmitted at all; whereas with parthenogenesis, you are certain to be in every offspring. This rule of inheritance that so strongly favors the option of parthenogenesis is called *the cost of sex*, although according to circumstance it might more aptly be called "the cost of tolerating cheap sperm cells," or of "having lazy males" (strictly, of allowing there to be fathers who don't do their share in child raising—males are far from lazy in the other things that they have to do).

If, on the contrary, you perceive that your bearer had a bad time, was not well adapted, and only just made it to being able to reproduce at all—or else if, although your own life was good, you perceive the environment to be in the process of very rapid and unpredictable change—then you might opt for sex on the ground that retaining your present dubiously serviceable gene combinations (that is, retaining the allies in your current genome), may make your offspring quite hopeless. then you might well reckon that the mere half chance of transmission due to sharing with a mate is more than offset by the chance of making up new random alliances that will turn out well. The object is these better new combinations—the mixture of some of your present allies along with some of someone else's. It doesn't greatly matter which happen to mix with which; or at least it does matter, but you can't predict what will be good, and random chiasma formation will see to it that you can't choose. (Actually this is not quite true; as a gene you can't choose anything, of course, certainly not any particular combination, but in the ensemble that is your bearer there may be the power of deciding what is a better bet. For example, the best may well be nearly the same as was best in the present generation. And the female you are in may have some power of detecting the males more likely to slant your offspring the right way—to carry genes that would be good to join or to replace your current allies: but this *sexual selection* story would take too much time to tell.)

For sex itself, arguing that things are *so* bad *so* often that union with a mate is the best course to take is not going to be at all easy. It has to be remembered that in the effort to make up better new combinations by sex, worse ones—and more of them—are also bound to be produced. This factor is offset somewhat if some elimination of hopeless offspring can occur very early, when not much biomass is lost and when they can be replaced by siblings. The effect of sibling competition in improving models of sex has long been recognized.[8] But not all highly sexual groups have such competition (for example, fish or crabs having tiny planktonic larvae), and without this factor the situation still looks rather bleak. Many who have tried seemingly hopeful assumptions in models have found how hard is the task. As I came to realize it myself, I became more and more pessimistic that the generally considered factors of evolutionary change could be strong enough or of the right kind. Constant strong pressure for a set course of change certainly will not do; the pressures themselves need to be changing all the time. They must change rapidly and promote now this and now that gene combination, and ideally it should all be done in a fairly erratic sequence so that no *regular* sequence of polymorphism (such as is produced out of a single genotype in a clone of aphids or water fleas throughout the seasons of the year) could succeed. Nor is it enough if, although unpredictable, the changes occur smoothly over a span of many generations. Then it is still going to be advantageous at least in the short term to stay with a known successful genotype. But if the long term is considered, might sex eventually succeed nevertheless? The issues at this point become rather subtle.

It could well be that, if parthenogenetic, a population can't keep up with change for any great length of time, because mutations adapting to the change do not occur fast enough. This will particularly be the case if needed adaptations involve large, carefully constructed molecules doing functional tasks: the loss of the gene for any such molecule because it is temporarily not needed may be the gain of a dormant seed of long–term disaster. Thinking of yourself as a sex–deciding gene again, it may seem that *if* you could be farsighted, you would opt for sex *even though* for some generations to come you would be come increasingly overwhelmed by the numbers of competitors who show parthenogenesis. The question is, then, whether the short–term numerical advantage of the clones becomes so great that you, the sexual determiner, go extinct while still waiting for your farsightedness to pay off. The chance of extinction makes it unclear if you will survive if the pace of change is slow, but equally unclear is the ability of asexual clones to survive in the long term.[9] The rather unexpected implication of the two ideas just stated, taken together, is that a form of group selection may have to play a part in explanations for the evolution and maintenance of sex. On the one hand, we can try to find factors in nature that are very rapidly changing (with diseases as one obvious good bet), and try to construct models that achieve the difficult task of paying the full cost of sex in the short term.[10]. Or we can accept an argument that there has been a very long history of population and species extinctions and this has borne hard on all units that were able to go asexual too easily, leaving only the units which, while they were sexual, had accumulated by accident (perhaps as side effects of genes selected for other reasons), various inbuilt barriers to the easy acquisiton of parthenogenesis.

Such barriers do seem to exist. For example, entry of sperm seems to be essential for further development of ova in many groups, leading to a situation where parthenogenesis, if it exists at all, is still disguised in lineaments of sex. Females of so–called "gynogenetic" asexual strains avoid the dilution involved in getting genes from male contributors, but their ova still have to be penetrated by a sperm before they can develop. An important point about this system is that the clones cannot exterminate the sexual strain that supplies the sperm without immediately going extinct themselves; and in fact the system seems always to set up an uneasy equilibrium of the two strains even though this is not an outcome of the most obvious theoretical dynamics.[11]

In so far as barriers to a quick, efficient start to parthenogenesis exist, we can accept progressively longer and longer–term environmental changes as supporting sex. During any phase of a cycle that lasted many generations, there might be rounds of selection towards improving a particular mutation to parthenogenesis because it is linked in a currently excellent and increasing genotype; but if these runs only start work on a process that is very inefficient–one, say, with only little over half of the parthenogenetic embryos surviving—there will not be enough time for progress to reach a point where the clone could drive the sexual strain extinct before, with changing phase of the cycle, the sexual's continued recombination began to pay off. Then it would be the imperfect asexuals that went extinct, and after that there can only be another equally inefficient start repeated during another phase.[12]

However, quite a lot of organisms—especially plants, but also lower animals at least up to the level of arthropods—seem to have easily available efficient parthenogenesis. This judgment is based on how often strains lacking males have been established and from the frequency of claims of having found examples of facultative parthenogenesis. Clearly the problem of short–term maintenance remains quite acute. Therefore let us now continue our search for sources of rapid and severe environmental change that might be able to immediately justify the twofold cost of sex. We aim to put any candidate process through a severe test: if we can find one that kills the twofold dragon, it will almost certainly kill the lesser (perhaps more realistic) dragons quite easily.

It might be thought that, even if the annual climatic cycle (as it applies to very short–lived organisms, like water fleas) is too regular for sex, there are plenty of unpredictable longer–period climatic changes. These changes push back and forth, but mostly are not regular enough to be called cycles.[13] At the long extreme, an example would be the ice ages. The trouble with them is that although they apply on a scale of generations to both shorter- and longer-lived organisms, and although they may have the right degree of autocorrelation (varying from near lack of it to a good periodicity as in the case of, say, sunspot cycles), the fluctuations are too unresponsive to the states of the populations they affect. Their extreme runs sooner or later drive all polymorphisms to fixation[14]—that is, one by one all the alleles go extinct until only one at each locus is left.

Might polymorphism, in spite of such runs of selection, be maintained by recurrent mutation? It might, but it is not very likely to work as a help for sex for two reasons. It has been suggested already that it isn't likely that simple remutations to gene states that code elaborate functional molecules will occur. Second, even if such mutations were somehow numerous enough, they assist asexual clones to cope with change just as much as they help the sexuals. Once polymorphism is truly lost, there is of course nothing that sex can do to make new combinations, and all possibility of advantage is gone. What seems to be needed is a more relenting set of pressures that ease off when alleles are near to being lost. This will tend to ensure that there is always something with which recombination can work.

It is a well-known principle in ecology that common organisms tend to attract many exploiters. This happens both within lifetimes of particular populations of the organism, and over long evolutionary time. If you look at the leaves of the commonest species of tree in your neighborhood, you are likely to see more of them damaged or infected than if you look at the leaves of a rare tree. This is the principle that makes the living world so varied. It implies that *biotic* pressures ease off when any species becomes rare. This slows its path towards extinction, and perhaps gives it a chance to await the opening of a new niche and a return to abundance. If the same principle can be applied to genetical variation within a species, then we see a possibility of a *biotic adversity* view of sex:[15] variation will be preserved in the species in the same way that species are preserved in an ecosystem. The crucial question, then, is whether such variation is largely static, or whether it is kept dynamic by some intrinsic ecological churning of the coadapting systems. If the latter is the case, there is hope that we are on the track of an answer to the puzzle of sex.

Before discussing possible dynamics, however, let us ask what kinds of enemies are most likely to maintain variation in a species. Enemies are basically of two kinds: predators who are big and eat whole individuals at a time, and parasites who are small and nibble or usurp nutrients—who damage but typically do not directly kill. The predators are almost always generalists and feed on many prey species. They are also normally longer–lived than their prey. I see them at present as being rather less likely to maintain an abundance of variation in prey, or to render the variation dynamic to the degree needed. This view may be mistaken. Some cycles of wildlife in the arctic, for example, have been attributed to a system of overshoots and crashes by populations of predators focused on a few normally abundant prey. During these cycles, the selection pressures on the prey certainly will change dramatically, and not only with respect to the predation itself. There might in these cases be a tendency to spare rare variants. But at present I cannot easily rationalize these possibilities as support for sex unless the other kind of biotic "pursuit"[16]—that is, parasite—is also involved. Parasites have a much more intimate approach to their food supply and are much more likely to differentiate between varieties of species. In fact, it is known that they often do differentiate; and because of the economic importance of parasites, in many cases the pattern of their discrimination has been worked out in detail.

"Parasite" is almost another way of saying "disease," so now I am back close to my title. A virus, an *E. coli*, a flea, a blood fluke, and a caterpillar are all parasites.[17] Some of them—the flea, the *E. coli*, the caterpillar—do not normally cause anything described as disease, but that does not mean that they exert negligible selection. The flea can transmit serious diseases; *E. coli* can turn aggressive when host defenses are low, indicating that it probably has to be under some restraint in our gut all of the time; and caterpillars do sometimes defoliate trees completely and kill them. Parasites, moreover, are ubiquitous; there is probably no sexual species that does not have several, and most have many with a fair proportion of their parasites devoted to the exploitation of that host alone. Parasites must undoubtedly have made their appearance at a very early stage of the evolution of life. I think this is one point on which the view of Lynn Margulis will agree with mine. Such immense age as a problem for hosts and such current ubiquity means that parasites, unlike most other factors that have been suggested, *have a potential to be the universal cause of sex.*

What pattern of interactions of parasites with their hosts is most likely to favor sex? Having decided this question, we may then further ask whether we find the specified pattern commonly in sexual species, and whether it is less common in species that have gone asexual.

One pattern of interaction that needs to be dismissed at once is overdominance within loci—that is to say, patterns where heterozygotes at each locus provide the fittest genotypes. Overdominance is the phenomenon usually illustrated by the well–known case of the sickle–cell anemia gene, which confers resistance to malaria in humans; and, in fact, that case is so well known that we have to beware not to assume all kinds of disease resistance are like it. Actually the pattern, at least with such strength of effects as is shown by the sickling gene, seems to be very unusual. I cannot afford space to describe the case and my objections to it as a paradigm of resistance, but will only note the following:

1. Such situations have stable population equilibria and do not lead to the *perpetuum mobile* that we require. The existence of such equilibria, however, makes like cases easier to detect and understand; thus perhaps the well–studied examples give an exaggerated impression of their importance.

2. A population would be considerably more fit if the heterozygote could breed true. Under sex, the heterozygote can't breed true; but after a duplication of the locus (which is not a rare evolutionary event), a situation equivalent to having a true–breeding heterozygote is easily established. Both loci would be homozygous (for different alleles) and there would be no contribution from sex. But parthenogenesis in the heterozygote would also achieve the same benefit and it would have the intrinsic efficiency of this mode of reproduction as a bonus. Thus heterozygote advantage may be good at keeping genes in play: but, like recurrent mutation, it only gives genes its protection in circumstances that allow even greater advantages for parthenogenesis.[18]

A more promising scheme for sex will have most the interaction of genes required for resistance not occurring within loci, as in the case just treated, but between them. A type of interaction that, like the last one, tends to preserve variability and protect alleles, is one where most genotypes have typically a rather uniform level of fitness, but all the time the environment, by temporally varying criteria, is singling out some genotypes as being extremely bad. This pattern preserves variability much better than a pattern where there are instead a few exceptionally good genotypes, while most are indifferent.

To visualize these two cases, consider the latter case first. We see sharp peaks of fitness standing up out of a uniform or undulating plain. The height of the peaks reflects the relative fitness of the genotypes that map to them. During times when there is little change in the environment, there is a rapid concentration of the population upon the peak genotype with accompanying rapid loss of alleles that happen to be represented only in the plain. To see how genetic variability is better preserved in the first option of few extremely bad genotypes, imagine deep pits in a flattish, coalfield–like landscape of fitness. There will be no strong tendency to lose alleles in this representation. On the theory already outlined, the parasite will concentrate on the commonest genotypes. Thus it is the common ones that tend to become pits. As a common genotype becomes a pit, it is the property of sex to be able to send many of the offspring of its outcrosses (that is, of matings other than with those with its own type) scrambling out of the pit. This coalfield–like landscape undulates over time, as did the previous landscape; but although pits may grow swiftly deep, like mine shafts, they are less likely to get anywhere near to engulfing the whole population (thereby driving many alleles to extinction) than is the selection by peaks.

A coalfield landscape of genetic fitness interactions can be argued to be a more likely outcome of parasite ecology and evolution than the peak landscape: peaks imply that some multilocus genotypes exist that resist almost all types of parasite attack. Considering the different natures of the various parasites that beset a species, this seems intrinsically unlikely. The coalfield, on the other hand, implies merely that whenever a genotype has grown common, all kinds of parasites evolve to become specialists at attacking it, causing the common genotypes to sink into a pit.

Such a verbal and pictoral argument is of course very far from showing that any given situation can be changing fast and radically enough for a twofold cost of sex to be met. Can it be worth such loss of efficiency just to have the chance of throwing some (certainly not all) of your offspring out onto the shoulders of an engulfing pit? Continuing in merely heuristic reasoning, there is at least a plausible line from ecology and behavior that may indicate how, for some species, the landscape may always be changing sufficiently fast to warrant the choice.

Many species have social behavior or other interactions that make small differences in phenotype magnify into much larger differences in expressed fitness. Two animals may be very close on a scale of size and health; and yet one of the two may achieve its full potential reproduction, while the other, just slightly below it on the scale, may achieve nothing. All animals of still lower ranks would likewise fail to reproduce. This pattern characterizes those higher species which are sometimes said to be more "K–selected," the term referring to their tendency to maintain set maximum levels of population density. However, a more appropriate term for them in the present context is that they are "soft–selected,"[19] this referring to the idea that selection is based on rank ordering, so that the achievement of a particular genotype varies greatly with the state of the population. Individuals in such species tend to compete face to face over limiting resources. In fact, they often seem to make more fuss about gaining control of these resources than one would have thought justified by necessity. The resource could be, for example, a territory or a nesting place within a colony area. Those animals that don't obtain any share in the contested resource become "floaters"; and as such, even though the resource they lack may not seem very crucial, their present reproduction is usually nil, and their chance of survival is also drastically reduced.[20]

If parasites affect health, and if even small differences in health affect the winning of contests, then such a pattern will markedly accentuate both the depth of pits in the fitness landscape and the rate of change of those pits. It will produce the rather flat landscape already described, whose main features are exceptionally deep holes; moreover, the holes will be filling and other opening all to a rather quick beat of the generations. With Robert Axelrod and Reiko Tanese, I have found in computer simulation that such a rank–based soft–selection scheme does make a model that can advance sex very effectively against the twofold cost. The models concerned are largely of a gene–for–gene nature (see below): but I have hopes that the same can be shown in a broader class, including ones where no subtle biochemical interaction has to be assumed to create the interlocus interaction. It is hoped that the interaction needed could arise from the stepwise fitness function implied above. In other words, if you are far enough up a rank order, for whatever reason, you get full fitness; if you are not far enough up, you get nothing. However, it has not yet been shown that this model, based on polygenic traits and this simple kind of interaction, will work.

Of course soft selection is often anything but "soft" for the organisms experiencing it. As already implied, soft does not mean weak, although there is no reason why soft selection should not sometimes be weak. It is called soft because there is no fixed relationship between a genotype and how many offspring it produces; instead this depends on the social situation—how many better individuals there may happen to be than a given one. There can be little doubt that something approaching this scheme does hold very widely for large long–lived organisms, especially animals. Such organisms are exactly the ones that can be expected to have exceptional problems with parasites, both because they are conspicuous to parasites and because parasites have such an advantage in adapting to them through the great discrepancy of generation times.[21] Satisfactorily for the theory, these large organisms are indeed the ones which are most consistently sexual. Small short–lived organisms tend on the contrary to be opportunistic. Their regime is more "hard selected"—that is, their reproductive performance is rather smoothly and rigidly linked to how well a particular genotype fits with its environment. A majority of these smaller species are actually parasites, or even parasites of parasites. Again corresponding with theory, the smaller and the more short–lived is the species, the more likely the species is to be asexual (although there are many exceptions).

It has been indicated above that the theory will work best if it is impossible to find genotypes resistant to all parasites at once, so in effect preventing the peaky landscape. Resistance instead should be rather a piecemeal affair, with many of the lines of adaptation mutually exclusive. Resistance has been so important to animal and plant breeders that there is actually quite a lot of evidence. Unfortunately I cannot state that, as a whole, it currently leans the way I would wish. Even in theory one can see ways in which whole sets of resistances could be strengthened simultaneously, and for the most part the evidence at present does not indicate mutual exclusivity for lines of adaptation.[22] For example, the immune system is a general purpose adaptation which aids higher animals in coping with the vast range of parasites—the "invention" of the immune system was in fact probably the crucial adaptation that permitted large homeotherms to evolve.[23] It is true that this system is very complex, and it is hard to imagine a single mutation which would affect all parts beneficially. Nevertheless, a gene that enlarged allocation to lymph nodes in embryogenesis, or one that enlarged the thymus, or one that enlarged bone marrow, could well have positive consequences for many facets of resistance at once.

Likewise, turning to large plants, trees, though they lack an immune system, could give the same sort of broad encouragement to resistance through genes that caused leaves to be more loaded with phenols or tannins, or simply to have more hard fiber. These attributes could confer a general resistance by deterring insect attack (although one needs caution here when the "deterrent" is not a poison).[24] For plants in general versus insects and fungi, a thicker cuticle might be of help.[25] Adaptations of such general nature are referred to as "horizontal" resistance by phytopathologists. What I have to admit is that descriptions of resistances of the horizontal kind seem to be more common, so far, in the literature of plant and animal breeding than are descriptions of highly specific resistance, and they are still more common relative to cases of which I have begun calling "complementary" resistance—that is, to cases where the genotype resistant to one pathogen proves highly susceptible to another and vice versa. The last pattern does occur,[26] and it may be that between diseases of very different types that are not often compared it is more common than is known. "Complementary" resistance is close to, although not identical with, the much discussed "vertical" or "gene–for–gene" resistances that phytopathologists contrast to the "horizontal" type discussed above. The latter seems to be particularly common between crop plants and their parasitic fungi, but some other parasites are involved and even some animal systems show it,[27] although here it appears to be much more rare. Even in plants it may not be so clear–cut a system as has been claimed.[28] Taken all together, then, the evidence for complementary resistance is scrappy; that for general resistance is much more substantial.

General resistance also must carry a cost of some sort. Otherwise every host species would have evolved to be resistant to every parasite, and it would follow on from this that all parasites would be extinct. This is *reductio ad absurdum* indeed, for the true situation couldn't possibly be more different: in species numbers, parasites rule the world. A parallel *reductio* is forced on us when we read the applied plant and animal breeding literature. Heritable resistance seems almost always to be found when it is looked for, and yet almost all the disease problems of crops and domestic animals continue. If anything, they grow worse or at least become a bigger facet of husbandry, due to the fact that as the variety of the genotypic bases of crops and breeds is reduced in response to the needs of mechanization, and through the possibility of the almost worldwide dissemination of particular successful varieties, the spread of pathogens is greatly facilitated.[29] Such a situation would be a complete puzzle unless there were a cost working in such a way that breeders are finally rejecting the most highly resistant lines because they are of low productivity;[30] or else are duly selecting a seemingly successful resistant strain and then finding problems coming from the complementary adaptations of an unthought of disease, which makes itself known soon after the new variety is planted,[31] rather as the Bad Fairy, forgotten at the party, comes remorselessly for her revenge on Sleeping Beauty.

If the former situation is the more common one (that is, it was horizontal resistance and it did have a cost), then *in nature* we would find that whenever a parasite epidemic ended, nonresistant genotypes would outbreed the others and susceptibility would come back. It is certainly possible to imagine a repeated cycle like this producing an advantage for sex, but the possibility has not been pursued. There might be cycles of abundance based on recurrent epidemics, and the epidemics could possibly be of quite different pathogens.[32] Suppose one cycle of epidemics has a different period compared to a cycle of some other kind affecting the species—say, that coming from pressure made on the host's resources by a second competing species that has its own cycle. If the genes for resistance to the plagues interact nonadditively with the genes that have to do with adaptation to competition, we have in essence a system that could support sex.

Nonadditivity might be very simple. It may not need to be anything more than the natural outcome of the rank–based soft–selection scheme already outlined. A sexual variant in this population might much more effectively track the changing requirements of its environment compared to a clone. Sex genes, hitchhiking along with changing instead of with fixed combinations could do well; and among sex alleles, those do best, perhaps, that promoted intermediate levels of recombination.[33] The level of recombination that evolved would depend, probably, on the periods and relative strengths of the cycles.[34] It has not been shown yet, but such a model might well pay the substantial cost of having sex. I regard the absence of models demonstrating how sex can be maintained in broad spectrum and polygenic resistance systems when treated along with reasonable opposed parasite ecologies, as one of the main current weaknesses of the theory.

In essence, under random mating, sex has one simple genetical effect: to release correlated associations of particular genes. It allows them to move towards more random association, and opens the possibility of quick formation of new correlated blocks. Under natural selection, some of the new blocks may be advantageous and become abundant (in diploids the new blocks also have opportunity to become more often homozygous); but it has to be emphasized that this later part of the process—selection causing changing abundance—has nothing to do with sex: given the same combinations to start with, asexuals would change as much or more. So looking both at the old combinations and the new after selecting, necessary signs to show that sex is not merely present but achieving something are (a) *nonzero linkage disequilibria* (this means that particular combinations of genes along chromosomes are unusually common—more common than could be explained by chance); and (b) *finding these linkage disequilibria are changing with time.* Are such nonrandom gene associations and changes in them known?

Unfortunately they are by no means yet known on the scale that my theory requires. Surveys of electrophoretic variation covering many gene loci do sometimes reveal linkage disequilibria, but they are usually quite small. I know of hardly any evidence of nonrandom associations that are changing with time. However, this latter fact may reflect more that lack of repeats of the surveys than a real absence of change. More generally, an excuse for the weak and absent effect involving linkage disequilibria can be that most of the enzyme systems studied in the surveys are definitely *not* those expected to be in the front line of defense against parasites.[35] On a more positive note, pointers to particular loci that may be connected with defense and disease are accumulating. These loci are not yet known to show any suitable nonrandom associations with each other or with other polymorphic loci, but then such association has not been sought.

The huge polymorphism of the Major Histocompatibility Complex is slowly revealing its connections with infectious disease. A truly functional role for its molecules in disease defense is at last beginning to appear: it seems that these molecules may actually hold things—pieces of virus or virus products[36]—up to the view of other cells as signs that the cell in whose membrane they are standing has been infected. On this view they do not (as in the earlier view, which was always strange to an evolutionist) simply signal the identity of their bearer. The finding of function for these molecules, taken together with the inevitable process of adaptation of microorganisms to avoid being caught up and revealed by the currently commonest histocompatibility antigens of defending hosts, is certainly a great help toward understanding the diversity at this locus.

The Major Histocompatibility Complex does show some nonrandom association of alleles at loci within itself, and doubtless sex has played a part to help these to be formed. However, linkages with the complex are too tight for them to be relevant to the short–term sex maintenance we are seeking at the moment. We need to look for other systems with which this one could be interacting.

Other loci with moderate polymorphism, but above all showing exceptionally high rates of evolution, have recently been revealed and discussed. These are the loci producing protease inhibitors in vertebrates.[37] They are as yet only conjectured to be connected with disease, but the idea is very plausible. The function of the inhibitors is to block the active sites of certain of the enzymes that digest proteins, so rendering those enzymes ineffective. My hopelessly unbiochemical imagination tries to sharpen its image of the interaction by seeing the product molecules as specially shaped logs that are thrown to jam the jaws of alligators as the latter swim toward a frail defending canoe . . . the canoe, of course, being the Vertebrate Body as it picks its way through a Swamp of Disease . . . In the real world, the conjectured interpretation is that since many pathogens, from bacteria to worms, are known to promote their attack by releasing proteases to digest host proteins, the protease inhibitors are a facet of the host's countermeasures. The "logs," in other words, have to be of just the right shape to jam the "teeth" of the alligators. Thus again, the Swamp of Disease keeps sending new alligators already bred to be different. So you too must keep changing the shape of your logs. Thus, in a kind of strip cartoon vision, I interpret the very high rates of evolution known to be occurring in loci, coding the contact sites on the inhibitor molecules.

So far, the rapid evolution disclosed for these molecules indicates high rates of gene replacement rather than a current polymorphism. The degree of actual polymorphism in the wild is not known, probably because it has not yet been a point of interest. The situation taken at face value prompts the thought that in general, mixed models having both dynamical polymorphism and rapid succession of complete gene transiencies (0 percent to 100 percent) seem a hopeful and an aesthetically acceptable solution for sex. The idea of numerous novel complete transiencies—new *universal* tooth patterns for alligators and new *universal* notches on logs—carries us back towards the old and currently rather discredited biological theory of sexuality, which said that sex existed because it permitted good new mutations to be brought together in single stocks with maximum speed. This idea was discredited partly because it was realized that it could not be imagined that new good mutations were occurring fast enough. Biologists were thinking then of genes like those that, since the Eocene, have been improving the adaptations of a horse's limbs for running, and good new genes in this sense do indeed seem to be very rare. But in the context of coevolution with microorganisms, there appears almost no limit to the potential demand for novelty— almost any new log with new shape may be better than nothing, while the Swamp of Disease is changing the teeth of its alligators very fast indeed. Bacteria have the potential to breed every twenty minutes in the warm, rich human body, were it not for the body's defenses. They get along very well with little sex, plenty of mutation, and a system of acquiring plasmids which to me seems rather like villagers hiring somewhat expensive and unruly samurai to help protect their enterprises. (Plasmids in a sense *are* sex for the bacteria, but that too is another story.) As for the evidence of rapid change in microbes, it is only too evident in the evolution of resistance to antibiotics by bacteria and in the constant transmutations of cold and influenza viruses.

Altogether it seems possible that there can be a fairly happy marriage between the "combining good genes" theme and the "dynamic multilocus polymorphism" theme that I have favored. A cycle of wide amplitude so that it shows fixations and remutations while essentially still going around the cycle may be a case that can be seen in the light of both the old theory and the new. However, I still very much doubt that such extremely long cycles as are implied in fixation–mutation successions can give short--term maintenance of sex, and so still expect to see true dynamic polymorphisms, that seldom fix and that carry strong nonrandom associations to and fro, playing their part.

If the idea that sex primarily defends against parasites is right, we would expect sex to be most universal where parasites bear heaviest, and to disappear where their pressure is relaxed. Inbreeding (which reduces effectiveness of sex) and parthenogenesis are both well known to be unusually common in organisms of extreme environments. This has been a puzzle in the past. It fits very badly, for exmple, with the idea that sex gets together good genes for overcoming vagaries of a harsh physical environment. On the other hand, it may turn out to fit very well with the idea that reduced biotic pressures in such environments cause reduced need for sexuality.[38] The sheer harshness of the arctic, high mountain slopes, edges of deserts, and disturbed agricultural land, where the selfers and parthenoforms are abundantly found, may by itself keep down parasites. Microbes, for example, may be too cold, too sterilized by sunlight, too desiccated, and so on, to be transmitted. But, in addition, parasites of all kinds may fail to coevolve along with their hosts *into* such habitats right at the start;[39] or, when they do establish, they may still fail to cross infect and contribute to the burden of other hosts of such localities because there are too few closely related potential host species on which they have any chance—a situation utterly different from what obtains, say, in old ecosystems in the tropics, where coexisting closely related species are very common. However, we should be careful here, because the plants and animals that succeed in living in the harsh environments may be extremely abundant there, and this may give those parasites that likewise succeed in reaching the habitat very easy opportunities of transmission. The reindeer tolerates a more arctic climate than any other deer; but, to judge from its polygynous system and its huge horns in the male (which are largest in proportion to the body for any deer), it is very insistently an outbreeding sexual animal. However, the reindeer is large and warm blooded, as well as both common and gregarious, and all of this, of course, makes it a very satisfactory host. Sure enough, parasites and diseases are present in abundance, and are an important factor in reindeer demography.[40]

Still in the arctic habitat, however, let us turn to grasses that the reindeer eat. These are often those most explicit of parthenogens that produce mitotically green seedlings growing out of their spikelets while the latter are still borne on the parent plant; and this and more cryptic types of asexual behavior and inbreeding in plants of the arctic are extremely common. Such perpetuation of unchanged genomes almost certainly would not succeed for the common species in a dense temperate meadow as some other dramatic recent evidence shows. Thanks to studies of grasses by Antonovics and his coworkers,[41] we begin to see not only that it is true that it does not succeed, but also hints that disease is the main reason. A recent experiment reared sexual and clonal offspring of sweet vernal grass (*Anthoxanthum*) and replanted them at the stage of rooted tillers in the meadow where the parents still grew. The results showed in various ways a very high advantage to the sexually produced offspring in their ability to thrive and produce their own seed. In the latest experiment, the performance of the sexuals is already, after two years observation, 1.55 times that of the asexuals. This is not quite the twofold factor, but since seedling mortality was not estimated due to the use of established tillers for the test, the real seed–to–parent factor may well exceed two; and in any case, the grass being perennial and the experimental plants still alive, the observed differences are likely to grow. These experiments are the only ones I know that show that sex may pay immediately, or almost immediately, its full cost. Since other influences bearing on success were carefully controlled for, and since damage by insects or even fungi was not apparent to a degree which could explain the difference, the investigators have suggested viral diseases to be a likely main cause.

It would be very interesting to know whether the same advantage of sexual offspring would have been found if *Anthoxanthum* had been studied at the very edge of its known range. Taking the life zones of the planet in a broad view, it is already established that tropical species have more of their gene loci polymorphic and also more general heterozygosity per locus.[42] In the case of some of the grasses ancestral to the cultivated cereals in Israel, it is also being found that marginal populations show both less disease incidence and less resistance to disease.[43] Taken together, such facts hint at a correlation of polymorphism with disease incidence. The pattern would no doubt partly be based on the density and type of spatial distribution, and partly on the levels of favorability of the environments to the transmission of disease propagules. It would be expected to hold both across ranges of species and within their ranges.

Obviously, much more needs to be known before we can say whether the ecological distribution of sex accords in detail with expectations of the parasite theory, and many more experiments of the type pioneered by Antonovics are also needed. Ideal for testing experimentally in a similar way would be short–lived, easily reared species that have both sexual and asexual races; or better still, those species which walk the wobbling tightrope of being able to use sex or not as they please. Such species might be made to evolve towards sexuality or asexuality by manipulating the extent of parasite attack. This has not yet been done, but a recent study by Lively[44] somewhat on these lines is almost equally well chosen. The study was of the distribution of sex in the wild in a New Zealand lake snail, and aimed to differentiate the present theory from several alternatives. Of all the factors reviewed by the author as potentially bearing on the snail's breeding system, only the extent of parasitism significantly predicted whether the snails would have more or fewer males in their populations. The fraction male was used as an index of the degree to which a population was behaving sexually. Unfortunately it is not known if the snails have mixtures of sexual and asexual strains or facultative asexuality; but either way, of course, the conclusion still accords well with the theory.

Using the obligately sexual mammals instead of the more undecided snails and grasses, Burt and Bell,[45] through a survey of the known chiasma cytogenetics of primates, found that only generation time was a good predictor of the rates of recombination. If the environment deteriorates at the same absolute rate for all species, and if recombination is a way of offsetting this deterioration as argued in the present paper (as also by Burt and Bell), then the positive correlation between generation time and rates of recombination is expected.

If this is a correct line to the explanation of chiasma frequencies, the involvement of age here forces to our notice an implicit point of the theory which has not yet been emphasized.[46] This is that the environment is becoming appreciably more threatening with respect to disease *within lifetimes of individuals.* Theoretical simulations of a population coevolving with its parasites, conducted by Robert Axelrod and myself, have confirmed the effect. They show not only that sexuals can beat asexuals in the short term when the parasites had much shorter lifetimes than the potential of their hosts, but have also shown increasing mortality rates within a lifetime. This change is due to the evolution of the parasites. Even though the reproductive and mortality properties of the hosts are set at the outset (that is, the hosts were potentially immortal), host mortality increases with age. The effect seen is slight, and in nature we would not sugggest it to be a major factor in shaping the senescent schedule of mortality; there are other factors more likely to produce this main effect.[47] However, perhaps here is one more reason why influenza epidemics seem to be particularly dangerous for the elderly. It is not just that the elderly are weaker due to their senescence; it may also be that, in help-ing to "passage" the pathogen—through their participation in epidemics ten, twenty, forty, or eighty years ago—the elderly have contributed most to its current adaptations: they may, in ef-fect have customized the virus to hit them.

So, new genotypes are always needed.

Or are they?

Will continuing strides in medical science make ancient sex tru-ly unnecessary? Will we have technological fixes, where for so long we gambled our genes in the way I have outlined? Will we at last be allowed to experience humankind's greatest dream and dread—growing old forever—doing so by a process of patching, pushing in new or replacement genes or their products from the outside?

On other fronts, technology seems to be striding faster still: soon we may be able to clone babies from cells of our own bodies. After that is possible, will we perhaps—more even than through the (to my eyes) dubious progress in genetic engineering—be fi-nally set free from the crazily inefficient and yet decorative, even risible, protective process that has dogged our line for a thousand million years?

I don't know the answer to any of these questions, but will give my own feeling. I certainly *wish* that the answer to all of them could be *"No"*; or at the least that it could be "Not for a long time." Speaking of the real danger to the species that might at-tend abandoning sex, rather than of my own romantic prefer-ences in the matter, I think that we should reflect that whether we name the giver God or chance, we may have been lucky in having AIDS as a first warning of things to come—warning that our own monoculture is already being planted too close, and that our own engines are making the winds that blow our own enemy spores around. Had the new plague bursting upon the massed jet–travel-ing human population of the second half of the twentieth century been a new virus caught in the manner of flu but still as lethal as AIDS, we should probably be already well aware of how fragile have been our victories so far over disease. With the factories un-manned, with the last stocks of antibiotics and vaccines failing, with insulin for diabetics used up, with blood and blood factors for hemophiliacs no longer collected, with blood dialysis machines unpowered, we should be finding little time to throw even consoling words to those *extra* dying—and those words the same barren ones, I suppose, that must be given to hemophiliacs dying with AIDS already: "Well, science gave you some additional years: we do hope that you have enjoyed them."

But this is only the beginning.

If such lethal flu strikes in the twenty–second century, when clonal man, or perhaps clonal woman, has become the human mode, with all of the even greater necessity for sterilized glassware and reagents and power supply, let us hope that such dangers of a sudden breakdown of civiliation have at least been well forseen and prepared form. For by then all people—*all*—may have within them the saved and "corrected" genes that are lethal in the absence of continual medical attention. If to be born again, I personally would prefer not to be born into the modish set of that time if there was any way not to be in it. I would prefer to be a remote, unlettered islander who had never seen even a penicillin pill, still less an implanted gene. Certainly I would not wish to be present with the mind and background that I now have, neither in my present body nor in my clonal rebirth, even that a neogerontology (or a technoembryology, or something of that kind) gives me the chance tomorrow. I am a primitive, I think: I was once told that I look like the last of the Neanderthals: perhaps because of this streak, I find I don't even want to be a part of a culture that considers it normal to have babies by caesarean section. Even that seems to me to be giving too many hostages to sterilized glassware and to chance. To this extent at least I am a believer in natural selection. Along with this belief goes a continuing quaint faith in sex—that it may for a long time do things for us that human planning can't.

This is not to say that I believe that the changes implicit in the set of questions I asked above won't happen, or that if they do they will necessarily be cut back by disaster, or that humans will go extinct: the currents are already plainly set, and most people seem to see none of the dangers that I do. And perhaps they are right. It may be that the building of the high–technology human modules of the future out of our increasingly (individually) hopeless bodies is as inevitable as was the building of the large long--lived multicellular creatures, once sex had allowed it to be possible. Of that long–past evolutionary building I am now an example. Had I been an independent protist flowing on the bottom of a pond, a crawler that had not yet even dared to adopt a spirochaete to be a flagellum,[48] would I have hung back on the threshold of *that* ancient revolution too? I suspect that I would, and doubtless would have looked foolish then as well. This paper is being poured hopefully, respectfully, as will be seen at the end, to modernize a gap between two sonnets of Shakespeare, to give him arguments for the future if he would need them;[49] but even he, I think, is probably more modern than I on the point I am worried about. At least the trend of *unnatural* human reproduction, for example, had begun long before his time. He knew about it and, to judge from some of his characters, it intrigued him also. His able, sturdy Macduff killed Macbeth, evading prophesy of the witches, by virtue that he (Macduff) was *"not of woman born,"* but was *"from his mother's womb untimely ripped."* The play hints that the mother survived this, for Macduff had brothers. Further, as is well known, so ripped was the more ambiguous Julius Caesar of another play. But in Shakespeare's time caesarean section had not begun its bid to become a normal method of human childbirth.[50]

What I want is not to reject medicine and the use, sometimes, of its extreme measures, but rather to say that I believe we should go carefully. We should try to understand not just the mechanisms by which our bodies work, but what might be described as the philosophy underlying that mechanism. We should try to understand the process that helped it to become as extraordinary on the intellectual side as it is—in which sex certainly has a great part. I want us to understand the whole chain of our being, that which connects us to the rest of living nature, not just the alloys and welding of special links—our own—and how to fix these. We need at least a theory of human psychical evolution that pays attention to the net of kinship behind us and still with us, to the effects of group competition in our makeup, to the extra innovations made by barter—and this to name but a few of the factors that are coming to be better understood[51]—before trying to grasp inflexible dogmas about what is right. Otherwise our attempts to change the world towards how we believe we wish it to be is almost certain to produce monsters. I feel that such evolutionary understanding will make us less cowardly about death and will deter the cant of claiming it is dogmatically and unequivocally right to cure disease, save lives, or fix infertility. The offspring which sex always dropped deeper into the mineshafts, those that it never expected to save—indeed that its very object was to eliminate efficiently[52]—should be borne in mind. We should reflect whether, in the longer run, technology has really any hope of saving such "selective deaths" even now. In the most immediate way, as dependent on my theme, I would like to see respect given for the views and rights of those who refuse any combination of medical intervention for themselves and their children. I find myself in this an improbable evolutionist ally to Jehovah's Witnesses against the Establishment of Medical Science. I think I look with eyes very like those of the Witnesses at some things—with horror, for example, at the white pall of hospitals that is rearing so high over our concrete and asphalt suburbs all over the world, seeing here the beginning of even greater arrogance and excess than when the churches of the Middle Ages spread—at least much more beautifully—over a then–greener face of Christendom.

Sex had done far more for us in evolution than simply to hold back disease—far more in fact than I have had time here to discuss. I would like to see sex kept not only for our recreation but also, for a long while, let it retain its old freedom and danger, still used for its old purposes.[53]

> O! none but unthrifts:—Dear my love, you know
> You had a father: Let your son say so.
>
> SHAKESPEARE, Sonnet XIII

NOTES

1. Bell, G. 1982. *The Masterpiece of Nature: The Evolution and Genetics of Sexuality.* University of California Press: Berkeley, CA.
2. Jaenike, J. 1978. An hypothesis to account for the maintenance of sex within populations. *Evolutionary Theory* 3:191–194. Bremermann, H. J. 1980. Sex and polymorphism as strategies in host–pathogen interactions. *Journal of Theoretical Biology* 87:671–702. Hamilton, W. D.. 1980. Sex versus non–sex versus parasite. *Oikos 35:282–290.*
3. *Anderson, R. M., and R. M. May. 1982. Population Biology of Infectious Diseases.* Dahlem Konferenzen, 1982. Springer Verlag: Berlin, Heidelberg, New York. In particular, the contribution of M.S. Pereira, 53–64, summarizes the current situation of human disease.
4. Eales, L. J., K. E. Nye, J. M. Parkin, J. N. Weber, S. M. Forster, J. R. W. Harris, and A. J. Pinching. 1987. Association of different allelic forms of group

specific component with susceptibility to and clinical manifestation of human immunodeficiency virus infection. *The Lancet* (May) 2: 999–1002.

5. This is only not true if the good trait tends to be dominant, which is claimed not the case in the example given. Basically, under sexual diploidy and random mating, selection works on means of samples of two (the two homologous genes). This is only half as efficient as when selection works on single genotypes, as is the case under parthenogenesis.

6. Grafen, A. In press. A centrosomal theory of the short–term evolutionary maintenance of sexual reproduction. *Journal of Theoretical Biology* 6.

7. When the option of asexual reproduction is completely open, as is implied here, in any period of environmental constancy the variant wastefully performing sex is likely to be lost completely, and extinction of the species may then follow when the environment again changes. This point will be discussed in more detail later.

8. Williams, G C. 1975. *Sex and Evolution.* Princeton University Press: Princeton, NJ. Maynard Smith, J. 1978. *The Evolution of Sex.* Cambridge University Press: Cambridge, England.

9. Tresiman, M. 1976. The evolution of sexual reproduction: a model which assumes individual selection. *Journal of Theoretical Biology* 60:421–431.

10. Hamilton, W .D. 1980. Sex versus non–sex versus parasite. *Oikos* 35:282–290.

11. Stenseth, N., L. R. Krikendall, and N. A. Moran. 1985. On the evolution of pseudogamy. *Evolution* 39:294–307.

12. Carson, H. L. 1967. Selection for parthenogenesis in *Drosophila mercatorum*. *Genetics* 55:157–161; Templeton, A. R., H. L. Carson, and C. F. Sing. 1976. The population genetics of parthenogenetic strains of *Drosophila mercatorum*. II. The capacity for parthenogenesis in a natural bisexual population. *Genetics* 82:527–542; Lamb, R. Y. and R. B. Willey. 1979. Are parthenogenetic and related bisexual insects equal in fertility? *Evolution* 33:774–775.

13. Examples of such selection are described in Van Norwijk, A. J., J. H. Van Balen, and W. Scharloo, 1980. Heritability of ecologically important traits in the Great Tit. *Ardea* 68:193–203; and in Grant, P. R. 1986. *Ecology and Evolution of Darwin's Finches.* Princeton University Press: Princeton, NJ.

14. Hamilton, W D., P. A. Henderson, and N. A. Moran. 1981. Fluctuation of environmental and coevolved antagonist polymorphism as factors in the maintenance of sex. In: *Natural Selection and Social Behavior.* R. D. Alexander and D. W. Tinkle, eds. Chiron: New York. 363–381.

15. Glesener, R R., and D. Tilman. 1978. Sexuality and the components of environmental uncertainty: Clues from geographical parthenogenesis in terrestrial animals. *The American Naturalist* 112:659–673.

16. Hamilton, W. D. 1986. Instability and cycling of two competing hosts with two parasites. In: *Evolutionary Processes and Theory.* S. Karlin and E. Nevo, eds. Academic Press: New York. 645–668.

17. Price, P. W. 1980. *Evolutionary Biology of Parasites.* Princeton University Press: Princeton, NJ.

18. Only half the offspring of a heterozygote through a sexual union are like the parent. Since by definition of overdominance the two homozygous classes are less fit, it cannot be advantageous to produce them. Thus parthenogenesis, which makes all offspring like the parent, is the best option even before its other advantages, accruing through the nonproduction of males, are taken into account.

19. Christiansen, F. B. 1975. Hard and soft selection in a subdivided population. *The American Naturalist* 109:11–16; Wills, C. 1978. Rank order selection is

capable of maintaining all genetic polymorphisms. *Genetics* 89:403–417.

20. See, for example, Jenkins, D., A. Watson, and G. R. Miller, 1964. Predation and red grouse populations. *Journal of Applied Ecology* 1:183–195; Smith, S. M. 1978. The underworld of a territorial sparrow: Adaptive strategies for floaters. *The American Naturalist* 112:571–582.

21. Hamilton, W. D. 1982. Pathogens as causes of genetic diversity in their host populations. In: *Population Biology of Infectious Diseases*. Dahlem Konferenzen 1982. R. M. Anderson and R. M. May, eds. Springer Verlag: Berlin, Heidlberg, New York. 269–296.

22. As one example of resistance to several taxa of pathogen, although all intracellular, see Skamene, E., P. Gras, A. Forget, P. A. L. Kingshorn, S. St. Charles, and B. A. Taylor. 1982. Genetic regulation of resistance to intracellular pathogens. *Nature* 297:506–509.

23. See note 21, above.

24. Hamilton, W. D., and N. A. Moran. 1980. Low nutritive quality as defense against herbivores. *Journal of Theoretical Biology* 86:247–254.

25. See, for example, Farnham, A. W., K. A. Lord, and R. M. Sawicki. 1965. Study of some of the mechanisms connected with resistance to diasinon and diazoxon in diazinon–resistant houseflies. *Journal of Insect Physiology* 11:1475–1488.

26. Some examples are in Hamilton, W. D. 1982. Pathogens as causes of genetic diversity in their host populations. In: *Population Biology of Infectious Diseases*. Dahlem Konferenzen 1982. R. M. Anderson and R. M. May, eds. Springer Verlag: Berlin, Heidelberg, New York. 269–296; and also Murty, B. R. 1971. Developmental traits in breeding for disease resistance in some cereals. In: *Proceedings of the International Atomic Energy Authority Panel Meeting on Mutation Breeding for Disease Resistance*. 12–16 October 1970. Vienna. IAEA Publication Number 271. 93–105. Both patterns in one system are shown in Bruzzese, E., and S. Hasan. 1986. The collection and selection in Europe of isolates of *Phragmidium violaceum* (Uredinales) pathogenic to species of European blackberry naturalized in Australia. *Annals of Applied biology* 108:527–533.

27. Gallun, R. L. 1977. Genetic basis of hessian fly epidemics. *Annals of the New York Academy of Sciences* 287:223–229; Parott, D. M. 1981. Evidence of gene–for–gene relationship between resistance gene H_1 from *Solanum tuberosum* ssp. *andigena* and a gene in *Globodera rostochiensis*. *Nematologica* 27:372–384; Rutter, J. M., M. R. Burrows, R. Sellwood, and R. A. Gibbons, 1975. A genetic basis for resistance to enteric disease caused by *E. coli*. *Nature* 257:135–136.

28. Barret, J. A. 1982. Plant–fungus symbiosis. In: *Coevolution*. D. Futuyma and M. Slatkin, eds. Sinauer: New York. 137–160.

29. Connor, S. 1986. Genes defend plant breeding. *New Scientist* 112:33–35.

30. See Murty, B. . 1970. *Breeding procedures in pearl millet (Pennisetum typhoides)*. Indian Council of Agricultural Research, New Delhi; for an effect of the same type, although not in an economic organism, see also Minchella, D. H., and P. T. Loverde. 1983. Laboratory comparison of the relative success of *Biomphalaria glabrata* stocks which are susceptible and insusceptible to infection with *Schistosoma mansoni*. *Parsitology* 86:335–344.

31. Murty, B. R. 1971. Developmental traits in breeding for disease resistance in some cereals. In: *Proceedings of the International Atomic Energy Authority Panel Meeting on Mutation Breeding for Disease Resistance*. 12–16 October 1970. Viena. IAEA Publication Number 271. 93–105.

32. See, for example, Canter, H. M., and J. W. G. Lund. 1969. The parasitism of planktonic desmids by fungi. *Osterreichische botanische Zeitschrift* 116:351-377. The paper concerns population dynamics without reference to genetics: however, genetic variation in resistance in phytoplankton is known.

33. Hamilton, W. D. 1980. Sex versus non-sex versus parasite. *Oikos* 35:282-290; Hamilton, W. D. 1982. Pathogens as causes of genetic diversity in their host populations. In: *Population Biology of Infectious Diseases*, Dahlem Konferenzen 1982. R. M. Anderson and R. M. May, eds. Springer Verlag: Berlin, Heidelberg, New York. 269-296; Bell, G., and J. Maynard Smith. 1987. Short-term selection for recombination among mutually antagonistic species. *Nature* 328:66-68.

34. Sasake, A., and Y. Iwasa. 1987. Optimal recombination rate in fluctuating environments. *Genetics* 115:377-388.

35. It might be asked here whether it isn't inevitable that some of the many gene loci surveyed will be at least linked to loci that *are* involved in defense. Won't the combinations of such linked genes be changed, if to lesser degree, along with the breaking and forming of the others that are really used in coadapting to the parasite? But the answer here is that if the frequencies of the marker genes that are seen are static, and if the population is large enough so that the effects of bottleneck combinations can be ignored, then such induced linkage disequilibria in neutral alleles will not occur. therefore I can remain hopeful that when the right genes are surveyed, fairly strong linkage disequilibria in the process of change will be found.

36. See, Self-examination. In: *Scientific American* 256:70; also Pollack, M. S., and R. R. Rich. 1985. The HLA complex and the pathogenesis of infectious disease. *Journal of Infectious Disease* 151:1-8; Smeraldi, R. S., G. Fabio, A. Lazzarin, N. B. Moroni, and C. Zanussi. 1986. HLA-associated susceptibility to acquired immunodeficiency syndrome in Italian patients with human-imunodeficiency-virus. *Lancet* November 22:1187-1189.

37. Hill, R. E., and N. D. Hastie. 1987. Accelerated evolution in the reactive centre regions of serine protease inhibitors. *Nature* 326:96-99; Laskowski, Jr., M., I. Kato, W. Ardelt, J. Cook, A. Denton, M. W. Empie, W. J. Kohr, S. J. Park, K. Parks, B. L. Schatzley, L. S. Oeyvind, M. Tashiro, G. Vichot, H. E. Whatley, A. Wieczorek, and M. Wieczorek. 1987. Ovomucoid third domains from 100 avian species: isolation, sequences and hypervariability of enzyme-inhibitor residues. *Biochemistry* 26:202-221.

38. Glesener, R. R., and D. Tilman. 1978. Sexuality and the components of environmental uncertainty: Clues from geographical parthenogenesis in terrestrial animals. *The American Naturalist* 112:659-673; Levin, D. A. 1975. pest pressure and recombination systems in plants. *The American Naturalist* 109:437-451. For the relative inability of asexual and selfing plants to withstand parasites, see Burdon, J. J., and D. R. Marshall. 1981. Biological control and the reproductive mode of weeds. *Journal of Applied Ecology* 18:649-658.

39. Reimers, E., L. Villmo, E. Gaare, V. Holthe, and T. Skogland. 1979. Status of *Rangifer* in Norway including Svalbard. In: *Proceedings of the 2nd International Reindeer/Caribou Symposium, Roros, Norway.* E. Reimers, E. Gaare, and S. Skjenneberg, eds. Direcktorat for vilt og Ferskvannsfisk: Trondheim, Norway.

40. Halvorsen, O. 1986. Epidemiology of reindeer parasites. *Parsitology Today* 2:334-339.

41. Kelly, S., J. Antonovics, and J. Schmitt. 1988. The evolution of sexual reproduction: a test of the short-term advantage hypothesis. *Nature*, in press.

42. Nevo, E. 1978. Genetic variation in natural populations: patterns and theory. *Theoretical Population Biology* 13:121–177.

43. For evidence of both fewer parasites and reduced resistance in marginal populations, see Moseman, J. G., E. Nevo, M. A. E. Morshidy, and D. Zohary. 1984. Resistance of *Triticum dicoccoides* to infection with *Erysiphe graminis tritici. Euphytica* 33:41–47; Segal, A., J. Manisterski, G. Fishbeck, and G. Wahl. 1980. How populations defned themselves in natural ecosystems. In: *Plant Disease: An Advanced Treatise.* J. G. Harsfall and E. B. Cowling, eds. Volume 5. Academic Press: London, England. 76–102.

44. Lively, C. M. 1987. Evidence from a New Zealand snail for the maintenance of sex by parasitism. *Nature* 328:519–521.

45. Burt, A., and G. Bell, 1987. Mammalian chiasma frequencies as a test of two theories of recombination. *Nature* 326:803–805. For similar facts for plants, see Stebbins, G. L. 1958. Longevity, habitat and release of genetic variability in the higher plants. *Cold Spring Harbor Symposia in Quantitative Biology* 23:365–378.

46. This point is especially important to an allied version of the theory of dependence of sex on parasites which attributes importance to very local deterioration of environment for each genotype, with sex protecting offspring from the parasite accumulated by parents and neighbors. This version is emphasized in Bremermann, H. J. 1980. Sex and polymorphism as strategies in host-–pathogen interactions. *Journal of Theoretical Biology* 87:671–702; Tooby, J. 1975. Pathogens, polymorphism and the evolution of sex. *Journal of Theoretical Biology* 97:557–576; Rice, W. R. 1983. Sexual reproduction: an adaptation reducing parent–offspring contagion. *Evolution* 37:1317–1320.

47. Hamilton, W. D. 1966. The moulding senescence by natural selection. *Journal of Theoretical Biology* 12:12–45; Bell, G. 1984. Evolutionary and non–evolutionary theories of senescence. *The American Naturalist* 124:600–603; Mueller, L. D. 1987. Evolution accelerated senescence in laboratory populations of *Drosophila. Proceedings of the National Academy of Sciences of the U.S.A.* 84: 1974–1977.

48. Margulis, L. 1981. *Symbiosis and Cell Evolution.* W. H. Freeman and Company: San Francisco, CA.

49. The question is why men should continue to exist, a point that may become as unobvious in the eyes of women as it is already in the eyes of biologists. Shakespeare's almost undivided theme in his first seventeen sonnets was to persuade an admired male friend that he should reproduce his beauty in offspring before he no longer can or dies. His verse emphasizes both male beauty and male heredity. The ending of Sonnet XI, for example, says *She* [Nature] *carv'd thee for her seal, and meant thereby* / *Thou shouldst print more, nor let that copy die.* In the modern future world, for peaceful life as an author and chance of tenure in a good department, Shakespeare had better not assume that male–male parthenogenesis is either the most desireable or the easiest to achieve.

50. Caesarean birth, although unrelated to the theme of infectious disease, is a good paradigm of the coming dilemma. As evolutionary background, it needs to be noted that the broadening of the female human pelvis, compared to other primates and the human male, to accommodate increasing fetal head size, clearly shows that substantial mortality of mothers in childbirth and/or a mortality or disability of birth–stressed offspring has been occurring. Accompanying the change, the shape of the pelvis has become a major constraint on women's average atheletic ability and was perhaps the main factor underling other trends of human sex difference occurring in the paleolithic. These dif-

ferences are usually regarded as having been favorable to men, although this seems to me possibly more a fashion of attitude than real. If we do decide that reducing the differences is desirable, it will certainly be easier to do if Caesarean birth becomes universal. But eventual costs of the new birth system will be: (a) necessity for medical facilities and personnel to be present at *every* birth; and (b) as the pelvis reconverges and/or fetal head size further increases, a prospect of *no going back.*

The increasing medical dependence implied in this example already has a momentum in public attitudes and therefore a kind of inevitablity. Should we *not* want this to continue, it is already hard to see what should be done. Mothers' lives obviously must be saved by Caesareans when necessary; but perhaps operations that are done in true necessity could be followed by an offer of a state reward for a pledge by the woman not to bear more children.

51. Alexander, R. D. 1987. *The biology of Moral Systems.* Aldine de Gruyter: New York.

52. I owe this stronger view to S. Nee.

53. A paper essentially identical to this is to be published by the Santa Fe Institute.

PANEL DISCUSSION

AUDIENCE QUESTION: How does your conceptualization of the coalfield–like landscape change when your assumptions are changed to allow for several loci rather than just two traits? Can there be multiple fitness pits in such cases and still have sex maintained?

HAMILTON: I don't know if sex can be maintained with multiple fitness peaks. I have wondered about it a lot, and the answer is that I am not sure. I can make the model work in cases where I see a single fitness pit appearing, and I guess I haven't analyzed my computer runs enough to say whether some involve more than one pit. As to the first part of your question, the general trend is the more loci the better. The more diseases are being resisted and the more loci are involved in the resistance, the better sex is maintained.

MARGULIS: I have been wondering about the explanation you gave for parthenogenesis being more common in extreme environments and on the edges of the geographic ranges of species. Is your explantion applicable to plants as well as large animals?

HAMILTON: The phenomenon usually goes by the name of "geographic parthenogenesis" and was first recognized in the 1920s, I think. It occurs in both animals and plants. In general, you are much more likely to find a race that is parthenogenetic on the edge of the range of the species than in the center of the range. My explanation is that parasites do best in the area where the species is most dense (that is, the center of the range) and where the species has been established the longest. Parasites have a difficult time hanging on in thinly dispersed, marginal populations.

RAVEN: To a botanist, your explanation would seem to run counter to common knowledge of the densities and ranges of species. Marginal races are often, in fact, very dense, very well represented, and more uniformly distributed on the edges of their ranges than in the centers. You might think that if the parasites had the right "key" they might attack them more efficienty on the edge rather than in the scattered populations in the center.

The standard botanical explanation for the increase in outcrossing in the center of the geographic range of a species compared to the periphery is that the center of the range is thought to be a more favorable habitat. People talk about the outcrossing enabling the offspring to exploit differing aspects of the more favorable habitat with their differing genotypes. They are able to do this because the environment is more favorable, allowing greater variability. None of this sort of loose reasoning is inconsistent with your idea, but the point is that other explanations already exist. One would still have to determine what is the central driving force that produces the differences between the center and the edge of the geographic range of the species.

HAMILTON: I agree with most of that and don't quite know what to add. My point has been that disease needs to be emphasized more in these kinds of explanations. I think it has been a neglected part of ecology because you have to be a bit of a laboratory type and a specialist to be able to identify disease organisms, and this doesn't go well with being the "naturalist" type who like to be out in the field. My hope is that ecology is shortly going to see a new interdisciplinary interaction which brings in plant and animal pathology.

AUDIENCE QUESTION: Plants lack an immune system, but many are asexual. Do plants have fewer pathogens, or do they have other ways of protecting themselves?

HAMILTON: I have been quite puzzled by the lack of anything resembling the immune system in larger plants.

RAVEN: A surprising array of chemical strategies have been found in plants in the last five years, and many of these are only brought into play after the plant is attacked. There has even been the suggestion that plants may transmit chemical signals through the air once they have been attacked. The presence of sensors for these airborne signals suggests plants may be more complicated in this regard than previously imagined. So I think things analogous to immune systems are being found and that we just have not spent enough time looking for them.

AUDIENCE QUESTION: Just how much damage has medicine inflicted on the genetic reserves of humans by its attempts to keep genetically unfit individuals alive and producing children?

MAYNARD SMITH: Speaking as someone who would be blind and

useless in a hunter–gatherer society without my spectacles, I'm extremely glad that people are keeping people like me alive and even allowing us to reproduce.

HAMILTON: I'm all in favor of keeping John Maynard Smith alive, or others like him, but I think there is potentially a problem in that we keep getting better and better at treating more and more people with genetic defects. I think there is a possibility that we might build up a situation where every one of our descendants has several lethal genes which require medication all of the time. If there were to be some breakdown of civilization, an earthquake or other kind of crisis, such that the medication could not be provided, then we might have an unstable, escalating crisis in medical care. If everyone is a diabetic, for example, the operatives in factories making insulin had better not join in any general strike or they may end killing themselves together with the rest of us. In the midst of such a crisis, the only survivors would be some lucky people on some South Pacific island that never had medical attention and so were still competent. I do think that there is going to be a growing problem along these lines.

MAYNARD SMITH: Don't you think that in the time scale in which medical treatment is going to lead to an increase in the frequency of deleterious genes, that we are going to see techniques for actually changing the genes themselves (that is, eugenics)? If we can actually transform deleterious genes into beneficial genes in the germ line, then I see no reason why in a hundred year's time we wouldn't do just that. Maybe I am too optimistic, but I see that as a sensible response to your idea.

HAMILTON: Such technical eugenics would indeed seem desirable. However, techniques are not going to arrive by magic, nor are corrected genes going to insert themselves into embryos by magic. Perhaps we should think right away about being more open–minded about research on early human embryos than we are now if we want to go that route. Some may decide that such research is even less desirable than going along with old–fashioned natural selection.

QUESTIONS FOR FURTHER DISCUSSION

1. Hamilton talks primarily about large, long–lived organisms. Can his ideas be extended to sex in the microcosm?
2. Why does Hamilton require "soft selection" to maintain sex in his coalfield model?
3. What does Hamilton see as the function of sex? Does it differ from Maynard Smith's view?
4. Hamilton passes quite quickly over the idea of sexual selection. Would sexual selection assist or retard the evolution of counter–defenses by host organisms?
5. Look back over the contribution by Margulis. In what way do the approaches of Hamilton and Margulis differ?
6. Why is it, do you suppose, that liver fluke worms reproduce asexually in snails but sexually in sheep?
7. Hamilton stresses parasites as a force in maintaining variation within a species. Are the large predators being dismissed too quickly? Could large predators be more important than microscopic pathogens?

Sarah Blaffer Hrdy uses her field experiences with primates to focus attention on the legacy humans inherit by virtue of their primate past. Rather than starting with the assumption that humans are something to be considered apart from other animals, Hrdy tries to remove the self–agrandizement humans tend to incorporate into their "objective" views of human behavior.

As you read this contribution, be especially aware of the shift in emphasis from the origin and maintenance of sex to the natural history of sexual interactions. A new dimension of sex is being added: the influence of sex on social organization. Hrdy is concerned with the consequences of our heritage as sexually reproducing mammals and what we must "rise above" in order to distinguish ourselves from the rest of the primates. The conflict of interests between the sexes has profoundly shaped our evolution, as well as that of plants, insects, and all other sexual organisms.

Watch for signs of bias in both Hrdy's presentation and in the literature she reviews. Ask yourself how much of what is reported is a reflection of these biases. The presence of these biases are not necessarily bad, just human, as Hrdy tells us in Chapter 7.

The Editors

5. The Primate Origins of Human Sexuality

SARAH BLAFFER HRDY

"Women are really dreadfully complicated," muses a lovesick veterinarian in Alan Ayckbourn's comic trilogy *The Norman conquests*. "With other animals, well, the majority of them, they're either off heat or on heat. I probably should have been born a horse or something." Indeed, the vet's life might have been simpler. For horses, like most mammals, have circumscribed, often seasonal, period of sexual receptivity. Within the limits, the answer to "Will she?" or "Won't she?" is remarkably predictable, so long as the timing is right.[1]

During heat or "estrus," a female mammal frantically solicits males with alluring scents, postures, and behaviors that are altogether absent at other times. this libido is rigidly controlled by the ovarian hormones, especially estrogens. Indeed, the word estrogen, as well as the term estrus, derive from the Greek for gadfly. The image evoked is a vivid one: females driven to distraction by this urgent buzzing in their endocrine system.

A number of primates, although not all of them, resemble pigs and horses in this respect. Sexual receptivity is determined by the rise and fall of hormones in the course of the ovarian cycle. Receptivity can be so strictly confined that in some prosimian primates like the galago, the estrous period is measured in just hours; copulation outside of estrus is physically not possible, because the vulva is sealed by an epithelial membrane. Among the best-known primates, savanna baboons and common chimpanzees, breeding can potentially go on all year, but copulations are confined to about two weeks in the middle of each thirty-five day cycle. During estrus, the sensitive skin of the anogenital area responds to rising estrogen levels by producing massive red swellings. In both chimpanzees and baboons, sexual swellings are a reliable signal that ovulation is imminent (see fig. 1).

When anthropologists interested in human evolution first went out to the field to study the behavior of monkeys and apes, the large–bodied ground–dwelling chimps and baboons happened to be the species selected for detailed study. Compared to nocturnal primates or the many small species of arboreal rain forest monkeys, baboons and chimps were easy to study. They live in large multimale, multifemale groups or communities that resemble the "promiscuous hordes" that early anthropologists had thought were an important state in human evolution, and their habitats inthe forests and savannas of Africa were arguably similar to those in which our ancestors were thought to have evolved. Furthermore, in the case of chimps, there was considerable anatomical and biochemical evidence indicating that chimps and humans

Figure 1. A female Barbary macaque, *Macaca sylvanus*, advertises ovulation with large swellings in the perineal region. (Photo courtesy of Meredith Small.)

shared a common ancestor as recently as 5 million to 10 million years ago. So it happened that baboons and chimpanzees, two species with pronounced sexual swellings at midcycle, figured prominently as templates for the imagined lives of our early ancestors.

As anthropologists came to ponder the evident absence of sexual swellings in our own species, they asked the obvious question: why had estrus been "lost" in the line leading to our own species? Rather than garishly advertised, ovulation in women had come to be "concealed"—hidden not only from those around them, but also from women themselves.

In retrospect, it is clear that such inquiries took a great deal for granted. For one thing, evolutionary scenarios were built on the assumption that our ancestors had a breeding system similar to that of chimpanzees today. It was taken for granted that our prehominid ancestors had possessed sexual swellings and that they had only been sexually receptive during discrete midcycle periods. In fact, we are unlikely to ever know whether or not this was the case, since the circumstances allowing us to test propositions about the tissues of 4–million–year–old ancestors will not often arise. But as so often happens in the history of science, a certain set of assumptions were virtually preordained by the rather more or less happenstance sequence in which research was carried out and by the idiosyncratic intellectual backgrounds of those shaping consensus in the fields of paleontology, primatology, and ethology.

WHY WAS ESTRUS "LOST"? OVULATION "CONCEALED"?

Whereas most mammals are characterized by discrete periods of sexual receptivity advertised by distinctive scents, visual cues, and behaviors, human females are considered to be "continuously receptive." When, in the early 1960s, Desmond Morris posed the question of why the "naked ape" was the "sexiest primate of them all," he launched a cottage industry in speculation about this topic. Over a dozen different hypotheses have been proposed to explain the "lost of estrus" and other attributes thought to be uniquely human. These included womanly breasts, concealed ovulation, a propensity to engage in ventro–ventral (face–to––face) copulation, and the capacity for female orgasm. The majority of these hypotheses take as their starting point a prehominid creature characterized by discrete periods of sexual receptivity at midcycle that are advertised by conspicuous sexual swellings. Matings are assumed to be promiscuous, or else females alternate periods of consortship with a particular male with periods of promiscuous matings, just as chimpanzees are known to do today.[2] Given that any of several males might be the father of the chimpanzee infant, the near absence of paternal care in chimps (and by extension creatures like them) is thought to be adaptive for males in the face of considerable uncertainty about which male could possibly be the father.[3] Given these assumptions, the question becomes, "How is a quadrupedal, small–brained creature like a chimp transformed into a large–brained bipedal human who, instead of concentrating matings into a brief period when the sexual swelling are maximally swollen, copulates less frequently but throughout the cycle?" The conventional answer from several generations of ethologists and anthropologists[4] is that we introduce the uniquely sexy, continuously receptive, continuously attractive, and uniquely sensual human female.

The notion of the "pair bond," and with it increased certainty of paternity, are central to all these models. Loss of estrus is seen as an adaptation that extends the duration of female receptivity and thereby increases the period of time that the male finds it sexually rewarding to be around his mate; the prospect of sexual rewards conditions him to continue returning to their shared home base and to bring with him provender needed for the couple's slow–maturing offspring. Indeed, according to the most elaborate model, published by Owen Lovejoy in a 1981 issue of *Science*,[5] our ancestors originally became bipedal so that males could more readily lug these provisions back to camp and their waiting mates. Explanations for the loss of estrus have centered around male provisioning and a corollary of such investment: greater male certainty of paternity. With males monogamously mated, the potential was there for Darwinian selection to favor fathers who were interested in and would help rear their own offspring.[6] Because these assumptions fit well with Western cultural values and our notions of the "natural" way for humans to live, there was little incentive to question them.

The various published scenarios to account for the loss of estrus are all intriguing, and some of them are dazzling in their complexity.[7] Among the most sophisticated elaborations on the "pair bond" hypothesis described above is that suggested in 1979 by Richard Alexander and Katie Noonan.[8] Alexander and Noonan proposed that females were selected to extend periods of sexual receptivity in order to force any would–be progenitor into a long consortship so that he would be precluded from establishing sexual relationships with any other females. A male in such close attendance could be fairly certain that the offspring his mate produced were his own; she in turn could rely on him to help rear their young, since it would be in the male's own genetic interests to do so. The Alexander–Noonan hypothesis went so far as to specify conditions under which this model would apply. The model should apply whenever the need for male parental care was increasing, but where the existence of mixed–sex communities (as in contemporary chimpanzees) meant that females would not be totally inaccessible to other males (as they might have been if our ancestors had lived in gibbon–like monogamous pairs spaced in individual territories). This is only one of a series of models in this vein.[9]

Not all of the hypotheses to explain the loss of estrus specifically assume a chimp–like ancestor. Some, like Spuhler's "libidinous biped" hypothesis, focus on the necessity for higher levels of circulating androgens in the increasingly bipedal, long–distance traveler that prehominids were becoming.[10] Enhanced female libido and continuous receptivity was an incidental by–product of hormone levels selected for by the need for greater endurance. Similarly, Mascia Lees and her coworkers have suggested that enlarged breasts and continuous receptivity were by–products of selection for "estrogen–related fat storage."

Other hypotheses to explain the loss of estrus are reviewed elsewhere.[11] The main point for this discussion is that each of these models is highly speculative. Each explanation of the loss of estrus makes a series of assumptions about what our ancestors were like, followed by construction of a hypothetical series of events providing the selection pressure for this or that change. Selected facts and anecdotes are marshalled to support a plausible, even logically compelling series of events. But the fact remains that we are attempting to reconstruct events that happened between 3 million and 10 million years ago. Choosing between competing scenarios becomes a matter of personal taste. Questions arise as to whether or not it is possible to investigate such topics without leaving the world of science and passing irretrievably to the world of science fiction—informed sci–fi, but fiction nonetheless.

PRIMATE SEXUALITY IN COMPARATIVE PERSPECTIVE

When we consider our ancestors of 4 million years ago, what can we say for certain? Of *Australopithecus afarensis*, we can say they stood about three to four feet tall and walked bipedally, but probably spent considerable time in the trees as well. For both sexes, brain size (380 cc to 450 cc) was only slightly greater than that of chimpanzees (300 cc to 400 cc). Fruit was probably an important component of the diet; what meat they did eat may have been obtained from either scavenging or hunting, and one or both sexes may have procured it. There is no firm evidence to indicate that males were the primary meat–getters, although this is the case for both chimpanzees and contemporary hunter–gatherers; nor can we be certain whether or not such males provided meat for mothers or infants. Early hominids were almost certainly social, and probably had babies who were at birth somewhat more mature than the highly dependent human infants with which we are familiar. Even so, these infants must have depended on their mothers and probably were carried by her for at least a year.[12] Beyond these generalizations, there is simply not much we can say with confidence about the reproductive careers of these creatures.

Under the circumstances, the least speculative option for reconstructing the breeding systems of ancient primates is to examine the full extent of the evidence about breeding in living primates in order to extract certain general principles, an approach known as "comparative primatology." Although the comparative approach yields probabilities rather than definitive answers, it offers several major advantages: it reduces the biases introduced by each researcher's own preconceptions about the social lives of our ancestors, and it offers a certain measure of testability.

Set aside explanatory models that extrapolate from one particular species, such as a chimp, and consider the entirety of the evidence known for other primates. Ignore for the moment the special attributes of humans: our capacity for language and culture, our multifaceted potentials for intimacy, and our ability to make conscious and often quite idealistic decisions. View *Homo sapiens* as just one among some one hundred seventy-five extant primates. Two things stand out: first, there is enormous diversity in the breeding system of primates; second, humans are not so unique in their sexuality as generally supposed.

Imagine a scale in which species like galagos, with strictly circumscribed periods of sexual receptivity, are listed at one end; and species with more flexible, what I will call "situation–dependent" receptivity, at the other end (see table 1). Humans, with their capacity to copulate at any point in the menstrual cycle or during pregnancy, fall at the far end of the continuum, but they are scarcely out there all by themselves. Other primates are scattered all along the continuum. Some of these species, such as the rhesus macaque, are seasonal breeders that copulate only during specific months of the year. During those months, mating may be limited to the period around ovulation, or may occur throughout the cycle. Some, like chimpanzees and baboons, exhibit sexual swellings at midcycle; in the case of the pigmy chimpanzee, females remain swollen for more than 70 percent of the menstrual cycle.[13] Among some of the monogamous New World monkeys, including owl monkeys, tamarins, and marmosets, females may copulate on any cycle day, and there is no detectable sign at ovulation. Indeed, since these animals do not menstruate, it is even more difficult to estimate midcycle than it is in humans. Similarly, in the organutan there is no visible signal at ovulation, although shortly after conception the female's genitalia swell and there is a slight whitening.

Substantial variation may also characterize a single species. Whereas common chimpanzees in the wild are quite cyclical, in captivity they may mate daily.[14] The most frequently reported circumstance (other than cycle stage) that precipitates sexual receptivity and assertiveness is an encounter between a female and an unfamiliar male. In a few cases, females not previously exhibiting estrus behavior, and even females known to be pregnant, may actively begin to solicit copulations. Such incidents have been reported for captive patas monkeys, wild hanuman langurs, wild gelada baboons, and wild redtail monkeys.[15] Similarly, Ranka

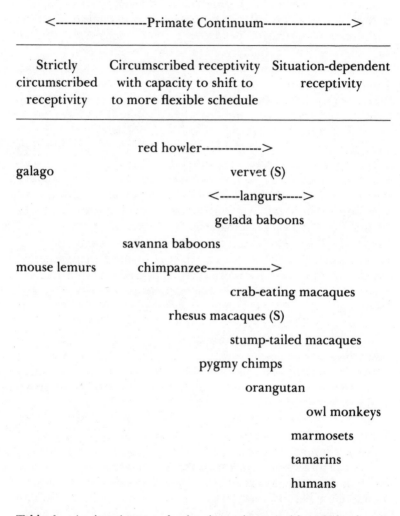

<----------------------Primate Continuum---------------------->

Strictly circumscribed receptivity	Circumscribed receptivity with capacity to shift to to more flexible schedule	Situation-dependent receptivity

red howler--------------->

galago   vervet (S)

<-----langurs----->

gelada baboons

savanna baboons

mouse lemurs   chimpanzee--------------->

crab-eating macaques

rhesus macaques (S)

stump-tailed macaques

pygmy chimps

orangutan

owl monkeys

marmosets

tamarins

humans

Table 1. An imaginary scale showing primates with strictly circumscribed, traditional estrus at one end, and primates with more flexible receptivity at the other. "S" denotes a strictly seasonal breeder.

Sekulic has reported an episode among south American red howler monkeys in which a solitary female became sexually assertive and receptive whenever her travel route brought her into the vicinity of males.[17]

Not all noncyclical receptivity can be attributed to encounters with strange males. Among seasonally breeding species as diverse as African vervet monkeys, Sumatran long–tailed macaques, and Japanese macaques, long bouts or "runs" of mating behavior with resident troop males can last for weeks or even months. Marmosets, tamarins, and owl monkeys are continuously able to copulate within the context of stable monogamous (or, as in the case of tamarins) occasionally polyandrous (one female, several males) groups.[17]

In virtually all monkeys and apes for which we have data, female sexual solicitations and female willingness to mate do show some tendency to peak at midcycle. Some swelling or at least slight puffiness as well as redness in the genital area are also common. Female attractiveness may also peak at midcycle—though, as in the case of the marmosets mentioned above, it need not do so. Of interest here is that the degree to which sexual activity tracks the female ovarian cycle may be influenced by which sex has the most control over the timing of copulations. Among captive orangutans for example, Ronald Nadler has shown that when the male takes the initiative, mating may occur on any cycle day in which a female is introduced into his cage. However, when orangutans were kept in special cages designed so that only the female (who is smaller in size) could pass freely from one cage to the other, there was a clear peak in sexual activity when the female was at midcycle. Nadler obtained similar results with chimps by training females, but not males, to use a door–opening apparatus. Female control resulted in greater cyclicity of sexual behavior.

Nevertheless, many of these species that exhibit peaks in sexual activity at midcycle also exhibit a *capacity* to engage in sexual activity at other times as well. In this respect, humans are not then so unique as widely supposed. Two examples serve to illustrate this point. The Japanese primatologist Y. Takahata recorded the reproductive biology of fifty–two free-ranging Japanese macaque females. During the October to March breeding season, females were sexually receptive an average of fourteen days, with a range of from one to ninety–two. Mounting behaviors during estrus were both heterosexual and homosexual, involving both female-–female and male–female mounting. Maria van Noordwijk describes similar patterns of sexual behavior for wild Sumatran long–tailed macaques. During the breeding season, females displayed no regular cycles, and were sexually receptive for as long as five to six weeks. During bouts of sexual activity, females would engage in sequential consortships with multiple male partners.[18] In both species, periods of sexual activity were interspersed with weeks or months when no sexual activity occurred at all.

A number of primates, then, are far more flexible in their mating habits than previously supposed. At the same time, it is important to note that humans fall significantly short of the high levels of sexual activity that the term "continuously receptive" would connote. As Frank Beach put it years ago, "No human female is constantly receptive" and "any male who entertains this illusion must be a very old man with a short memory or a very young man due for bitter disappointment." For, in addition to endocrine factors, highly subjective elements such as social conditions, cultural norms, partner preferences, and "mood" contribute to a woman's level of sexual arousal. This is one reason why I prefer the term "situation dependent."

The extraordinary variability in the patterning of sexual behavior among human females is by now well documented.[19] Nevertheless, in studies that focused solely on sexual behavior initiated by females, the results have been fairly consistent. In populations as different as a college community in Connecticut, a Kung San gathering and hunting group in the Kalahari desert, and a sample of American lesbians (a group presumably uncomplicated by any male involvement), researchers report a peak in sexual activity at midcycle.[20] This midcycle peak was sometimes also accompanied by a second, minor peak just prior to menstruation; interestingly, this resembles that reported for rhesus macaques.[21]

Assuming that this convergence between the patterning of sexual activity found among some women and the midcycle peak in sexual solicitations that characterize other primates is real, how would we characterize the difference? Women appear to be constantly "copulable" (to use Frank Beach's term), but female sexual assertiveness or libido ("proceptivity" is the technical term) appears to be moderately cyclical. Outstanding differences between humans and other primates include the strong role played by culture in human affairs and the relative lack of seasonality in humans compared to species like macaques.[22] In addition to such seasonality, many (but again, not all) primates undergo long periods of nearly complete sexual abstinence during lactation.[23] In terms of the patterning of their sexual behavior, then, a comparative perspective blurs some of the long–held criteria supposed to separate humans from other primates.

REVISING OLD QUESTIONS

More accurate information about the patterning of sexual activity in other primates invalidates the old notion that there exists a strict dichotomy between "continuously receptive" humans and nonhuman primates which are only sexually receptive during strictly circumscribed periods of estrus. There are a variety of semisolitary (orangutans), monogamous (marmosets, tamarins, and owl monkeys), single–male polygynous species (what we used to call "harems"), as well as at least one multimale polygynous breeder (vervets), where mating outside of midcycle is possible or even common. Furthermore ovulation may be conspicuously advertised (as in barbary macaques), concealed (as in marmosets), or else advertised over so long a period (as in pygmy chimps) that ovulation is for all practical purposes obscured.

The traditional anthropological question, "Why do humans conceal ovulation?" presumes that our prehominid ancestors did not do so, and that our recent forebears were unusual in doing so. In fact, our ancestors may or may not have advertised ovulation. Furthermore, truly continuous receptivity does not exist anywhere, and the flexible or situation–dependent receptivity that does characterize modern humans is far from unique. Under the circumstances, it does not make much sense to ask, "Why was estrus lost?" Such a question is not only unanswerable, but unnecessary. A comparative perspective prompts us to ask instead: "Under what circumstances do primates generally shift from cyclical to more situation–dependent receptivity? Under what circumstances do we find conspicuous signaling of receptivity? When is it absent?" In contrast to hypotheses that hinge on knowing about the sex lives of fossils, these new questions lead to hypotheses that are actually testable within a scientific framework.

WHY HAVE SEXUAL SWELLINGS?

Consider the question of sexual swellings. In most primates. female morphology and coloration change with different phases of the menstrual cycle. Such changes may range from very slight pinkening of the vulva at midcycle to enormous red swellings in the perineal area conspicuous at considerable distance. Of some 175 odd species of primates, approximately twenty–four exhibit conspicuous perineal swelling at midcycle (see table 2). Although it is clear that the swellings play a role in signaling female sexual receptivity and attracting males, there is no simple relationship between size of swelling and attractiveness. Indeed adolescent females, who tend to be less attractive to socially dominant males, often exhibit the largest swellings.[24] Other factors, including the identity of the female, past history of her relationship with the male, and olfactory cues, are also important.

Sexual swellings have apparently evolved more than once among the prosimians, ceropithecine monkeys, colobine monkeys, and great apes.[25] Most of the species that exhibit swellings live in multimale social units.[26] Among the two dozen species in the Colobine subfamily, for example, only three are characterized by breeding systems in which troops typically contain more than one breeding male on a permanent basis. Interestingly, these are the only three species in the subfamily which are characterized by prominent sexual swellings (the red colobus, olive colobus, and black colobus). Similarly, in the superfamily Hominoidea, only common chimpanzees and pygmy chimpanzees live in multimale troops, and only these species exhibit sexual swellings.

There are several exceptions to the association between sexual swellings and multimale breeding systems. Prominent among them are such multimale species without swellings as vervets and rhesus macaques, and unimale species with swellings such as hamadryas baboons. Nevertheless, the association has been strong enough to entice researchers to construct a number of hypotheses to explain how females in multimale breeding systems benefit from swellings. For example, it has been suggested that by advertising impending ovulation, sexual swellings function to incite

Prosimians
  tarsier   either solitary or monogamous

Old World Monkeys	*multimale*	*unimale*	*unkown*
red colobus	x		
olive colobus			x
black colobus			x
barbary macaque	x		
liontailed macaque	x		
pigtailed macaque	x		
moor macaque	x		
Celebes macaque			x
Formosan macaque			x
Talapoin monkey	x		
Allen's swamp monkey			x
grey-cheeked manabeys	x		
white-collared manabeys	x		
sooty manabeys	x		
savanna baboons (3 species)	x		
guinea baboons	x		
hamadryas baboon		x	
drill	unimale in multimale group		
mandrill	unimale in multimale group		

Great Apes
| common chimpanzee | x | | |
| pygmy chimpanzee | | | x |

Table 2. Primate species exhibiting conspicuous sexual swellings visible to humans (and presumably other primates) at considerable distance.

competition between males for sexual access to females, thereby increasing the likelihood that only the "best" male will be able to monopolize the female at the critical time and hence father her offspring. The main alternative to this "single best male hypothesis" is that sexual swellings function to increase the female's chances of mating with a "range of male partners."[27]

WHY HAVE MULTIPLE PARTNERS?

The phenomenon of multiple matings by females is poorly understood. Indeed, even the existence of this behavior pattern was overlooked until recently. This may be partly due to Darwin himself. In setting down his theory of sexual selection in 1871, Darwin wrote, "It is shown by various facts, given hereafter, and by the results fairly attributable to sexual selection, that the female, though comparatively passive, generally exerts some choice and accepts one male in preference to the others."[28] Victorian stereotypes were then incorporated in evolutionary theory, and from Darwin onward it has often simply been assumed that female sexuality was directed towards insemination by *one* ideal male partner. Yet recent observations of a wide range of mammals, as well as longstanding descriptions of wild primates, make it clear that females mate with a range of male partners both at times when conception is possible and at times when it is not. Only recently has this widespread phenomenon been examined by theoreticians, and an extensive literature is emerging on this complex topic.[29].

At the risk of oversimplifying what I know to be a multifaceted phenomenon, I will focus here on selection pressures that I believe would be most relevant to the primate case. In virtually all primates, the behavior of males potentially has a critical effect—either positive or negative—on the survival of infants. Hence behaviors by females that affect subsequent treatment of their offspring by males will be a critical component of female fitness. If, as is being suggested here, males remember the identity of females with which they mate, and if they respond differentially to offspring of those females (a testable proposition), then there would be powerful selection pressures on females to mate with a range of male partners. Such males might be more tolerant of that female's offspring (in the sense that they would be less likely to attack the offspring), or they might be more likely to protect or care for them in situations where such behavior by males is an option. In fact, situations where males can invest in infants are probably more common among primates than we realized.

One of the misconceptions that has hampered our efforts to understand primate breeding systems was the conviction that high confidence of paternity was necessary before "paternalistic" or protective behavior by males could evolve. But, so long as care by males is relatively low cost and nonexclusive (spread out among several infants), there is nothing to prevent the evolution of low levels of male investment even in "promiscuous" or multimale mating systems with uncertain paternity.

To date, savanna baboons provide the best–documented example of males drawn into the protection of particular infants through their past (and in a few cases anticipated future) relationships with the infant's mother.[30] By mating with several different males, females forge a network of alliance with males. Even though only one male can be the biological father of the infant, several males nevertheless often assist a female by "babysitting" or protecting her infant, allowing the infant to forage close by, or moving to pick up the infant should it be threatened by strange immigrant males.[31] In species living in particularly harsh or dangerous environments, or where the costs of motherhood are particularly high, as among barbary macaques or tamarins, females mating with more than one male elicit far more than protection. Being carried about by males—typically males with whom the female had past consort relations—is essential for infant survival.[32]

By drawing several different males into the web of "possible paternity," females may increase the likelihood of male protection and even care. In addition, consortship with a male may decrease the likelihood that he would attack infants subsequently born to a female. Information on infanticide by adult males in some fifteen different species of primates belonging to such diverse genera as *Presbytis, Colobus, Alouatta, Cercopithecus, Pan, Gorilla,* and *Papio* make it clear that such attacks are most likely to occur when males enter a breeding system from outside—that is, when males find themselves in the company of mothers with infants that could not possibly be their own.[33]

There appear to be several different reasons why it would behoove a female primate to spread the possibility of paternity among several different candidates, any of whom might one day be in the position to affect the survival of an infant either by protecting or caring for it, or at least by not harming it. Instead of the old Darwinian assumption that the object of the female sexual signaling and solicitations is to mate with the single "best" male, a new possibility is introduced: among these primates, the female's goal is to mate with a range of males, regardless of which inseminates her.

Sexual swellings then would be exactly what they appear to be: striking advertisements of sexual receptivity that attract the attention of a number of available mates. The formation of multiple brief consortships is the object of the exercise, flamboyant sexual skin is simply an energetically efficient way of achieving it. Instead of having to take time away from foraging to solicit each male in turn, the female baboon or chimpanzee advertises to the community at large; the time and energy burden of finding an opportunity to mate is shifted onto the male.

This of course is from the female's point of view. From the male perspective, he confronts special challenges from living in a multimale breeding system where females routinely solicit multiple partners. Selection pressures on females to evolve sexual swellings and to evolve the sexual assertiveness that causes them to solicit multiple sexual partners have led to counter-adaptations among males—in particular, very large testes.[34] Zoologist Roger Short has proposed that testicle size should be correlated with sperm-making capacity, so that in species with multimale mating systems where each male mates sequentially with the same female, successful insemination will depend not only on the male's ability to gain access to the female, but also on the quantity and quality of his sperm relative to that of competitors. Among species like the chimpanzee, the contest for reproductive success is partially decided at the level of competition among sperm cells.

A comparison across species supports predictions generated by Short's model. The testes of a male chimp account for 0.27 percent of his body weight, compared to 0.06 or less in the case of other hominoids who live in either monogamous or one-male-/several female breeding units. Gorillas (who live in units where only one silverback male at a time breeds) weigh as much as 170 kilos—nearly four times more than most chimps—yet they have testes one–quarter as large. With few exceptions, testes are larger in the species with multimale breeding systems, smaller in the monogamous species such as gibbons or in the primarily unimale species such as langurs (see table 3). In macaques (which are not only multimale breeders, but also seasonal breeders where females concentrate the period of receptivity into a few months each year), selection pressure for large testes has been at its most intense. Even the apparent exceptions to Short's model have turned out to be revealing. Tamarins, which at the time were thought to be an obligately monogamous species, had surprisingly large testes. Now that we recognize the polyandrous potential in this species, the large testes are consistent with Short's hypothesis: if tamarin females were mating with several adult males in the group, sperm competition would indeed matter. In tamarins,

Species	Body weight (kg.)	Testes weight (g.)	Testes as % body weight
human	65.6	40.5	0.06
gorilla	169.0	29.6	0.02
orangutan	74.6	35.3	0.05
chimp	44.3	118.8	0.27 (multimale)
crab-eating macaque	4.4	35.2	0.80^s (multimale)
savanna baboon	24.3	52.0	0.21 (multimale)
squirrel monkey	0.78	3.2	0.41 (multimale)
langur	17.0	11.1	0.06
gibbon	5.5	5.0	0.10
tamarin	0.52	3.4	0.65

s = seasonal breeder

Table 3. Harcourt et al.* analyzed data on testes size relative to body weight for a wide range of primates. This table gives selected examples to illustrate the cases discusses in the text.

*Harcourt, A. H., P. .H. Harvey, S. G. Larson, and R. V. Short 1981. Testis weight body weight, and breeding systems in primates. *Nature* 293:55–57.

Figure 2. A langur female at midcycle solicits a male by presenting and shuddering her head, but there is no conspicuous morphological signal of ovulation. (Photo courtesy of Daniel Hrdy.)

it appears that large testes were a secondary adaptation, tracking the evolution of multimale mating systems.

So far I have explained why sexual swellings might be advantageous in some circumstances. Under what conditions might females benefit from not advertising their reproductive condition? Can "concealed ovulation" in humans be explained by using a comparative approach?

WHEN IS RECEPTIVITY SITUATION DEPENDENT INSTEAD OF CYCLICAL?

Much of the discussion of primate sexual behavior has focused on primates such as chimpanzees and baboons that live in large multimale troops, and who advertise ovulation. In fact, most primates do neither. Only a portion of primate species live in such groups. The majority live in other groupings ranging from mother–offspring units, to monogamous pairs, to situations where one male travels with a group of females. In some instances several males are present, but one male has priority over other residents in terms of breeding access; this exclusivity is often challenged by extra–troop males or by male bands, which temporarily enter the troop to mate (for example, gorillas, gelada baboons, langurs, patas, and blue monkeys). Finally, there is the newly discovered pattern of monogamy grading into polyandrous units reported for South American tamarins. Although I would argue that to some extent a "polyandrous component" is almost universal among primate breeding systems, so far tamarins comprise the only true case of polyandry in nonhuman primates comparable to systems of that name in birds and humans.[35]

Females in primarily unimale systems, such as langurs, exhibit midcycle peaks in sexual receptivity during which they mate with the resident or locally dominant male; but they retain the option to shift to more flexible, situation–dependent receptivity under certain conditions, such as when extra–group males invade the troop. Since such encounters occur on a nonpredictable basis, females might not be ovulating when the encounter occurs. Under such circumstances, females would scarcely benefit from advertising ovulation; indeed a female would sacrifice considerable flexibility in her mating options by doing so (see fig. 2). The same arguments might also apply to marmosets and owl monkey females that live in monogamous mating systems. In species with multimale breeding systems (for example, chimps, baboons, mangabeys), females compress their sexual activity into discrete, well–advertised intervals during which they mate with more or less all males present in the troop; once the brouhaha of breeding is done with, females return to business as usual, which is foraging. By contrast, females in one–male polygynous and monogamous breeding systems trade the efficiency of discrete estrous intervals for the greater flexibility in the timing of copulations. This flexibility is possible when ovulation is concealed and receptivity situation dependent. For these females, there would not be any selective advantage to midcycle swellings. Table 4 outlines a simplified version of this model to explain the distribution of primate patterns of hidden ovulation.

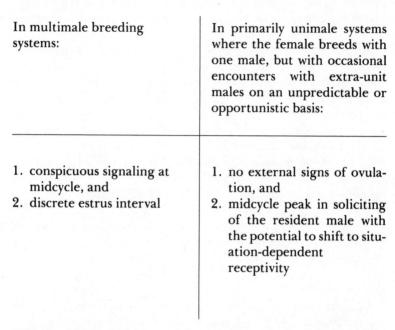

In multimale breeding systems:	In primarily unimale systems where the female breeds with one male, but with occasional encounters with extra-unit males on an unpredictable or opportunistic basis:
1. conspicuous signaling at midcycle, and 2. discrete estrus interval	1. no external signs of ovulation, and 2. midcycle peak in soliciting of the resident male with the potential to shift to situation-dependent receptivity

Table 4. This model assumes that monkey and ape females benefit from establishing consortships with a range of male partners. How to accomplish this becomes "the problem." The solution varies according to the type of mating system in which the female finds herself.

Although for many years the popular explanation for the absence of estrus in *Homo sapiens* has been the role of "continuous receptivity" in cementing pair bonds, it should be apparent that this is scarcely the only possible scenario. Sexual activity is remarkably uncommon in the most obligately monogamous of the primates, such as indri or gibbons (gibbons only come into estrus every two or three years and at that time confine breeding to a period of a few months), it is relatively much more common in those species such as tamarins, in which facultative monogamy grades into multimale (polyandrous) breeding systems. Not only is there a poor match between pair bonding and "continuous receptivity," but it is also now apparent that humans themselves are far from unique as regards flexible receptivity. Viewed from a comparative perspective, humans resemble a number of other primates that live in unimale mating systems where females occasionally encounter additional males on an opportunistic and unpredictable basis[36].

No doubt the sheer elaborateness of sexuality in the human case deserves some explanation. Given the greater role that consciousness plays in human existence, however, it is scarcely surprising that sexual relations among humans have become linked to a wider range of emotional experience. Nevertheless the basic outlines of our flexible mode of reproduction—concealed ovulation and the absence of circumscribed estrus—do not necessarily require any special explanation.

RECONSTRUCTING ANCESTRAL MATING SYSTEMS

The idea that female hominids became continuously receptive in order to facilitate pair bonding and to encourage their mates to bring food back to camp is the most widely cited explanation for the sexual nature of the human species.[37] Several of the species with situation–dependent receptivity and concealed ovulation are indeed monogamous (for example, owl monkeys and marmosets), but many are not. Pair bonding is a very weak explanation for the absence of estrus in *Homo sapiens.* What about the other quintessentially human traits, such as womanly breasts, face–to–face copulations, and the capacity for female orgasm that the pair-–bond hypothesis is supposed to explain?

None of these attributes has been the subject of detailed comparative inquiry in the same way that testes have been. Nevertheless, even on the basis of a superficial comparison across species, it is apparent that there exists no justification for assuming that such traits must necessarily have evolved in the context of the nuclear family. Let us begin with prominent breasts, the only one in the above list of "uniquely human" traits that actually appears to be one. Although other primates develop hemispherical breasts and pendulous nipples during lactation, these gradually shrink back to a scarcely noticeable condition once the infant is weaned. By contrast, human females start to develop prominent breasts in adolescence, prior to childbirth. Human breasts, like those of monkeys, and apes, may expand during pregnancy and especially during lactation, and then decrease in size after weaning, but they remain a permanent attribute throughout life.

The most widely cited explanation for this peculiarity is that breasts were sexually selected to enhance a female's attractiveness to her mate (as in the pair–bond hypothesis discussed above). There are a number of competing hypotheses, however, including the suggestion that permanently enlarged breasts serve to signal to males that (a) the female's fat reserves are sufficient to support pregnancy and lactation; and (2) the idea that breasts are a by–product of natural selection (that is, selection for survival value rather than for attractiveness to males) for increased capacity to store fat.[38]

Although permanent enlarged breasts are unique to human primates, none of the other attributes are actually unique and often they are found in species that are not monogamous. Face–to-–face (or technically, ventro–ventral) copulations, as well as sexual foreplay of all sorts, are well documented for both wild and captive orangutans and pygmy chimpanzees.[39] Whereas orangutans are as solitary as a primate can be, pygmy chimps live in multimale, multifemale groups; neither species is truly monogamous.

Female orgasm is also not unique to humans. This peculiar psychophysiological response, or else something very like it, occurs in other primates; and many of those species where it is best documented are definitely not monogamous.[40] Furthermore, recent research by Masters and Johnson and others renders absurd a basic premise of the pair–bond hypothesis; the notion that the female orgasm evolved in order to "make it easier for a female to be satisfied by one male."[41] Based on both clinical observations and interviews with women, there is a disconcerting mismatch between a female capable of multiple sequential orgasms and a male partner typically capable of one climax per copulatory bout. Furthermore, only a minority of women (on the order of 30 percent) typically experience orgasms from intercourse alone. Even for natural selection—rarely an agent of perfection—this level of response seems peculiarly substandard to warrant the claim that the orgasm is an adaptation for fostering pair bonds—though this is not to say that the orgasm, once it became part of the primate repertoire, might not subsequently be enlisted as an agent to promote pair bonding when social or environmental conditions are conducive to monogamy.[42]

Those stressing the inefficiency of the female orgasm for cementing pair bonds have typically assumed that the females involved were mating with a single partner, and that climax should occur as the result of stimulation normally experienced from mating with one partner. Indeed, in the experimental studies on this, the female was often caged with a single male. Yet in the wild, these same chimpanzees, baboons, and macaques are mating sequentially with multiple partners, and stimulation from these encounters is cumulative. In the most extreme cases (barbary macaques would be an example), females may change partners every ten minutes or so and sometimes mate with as many as three different males in six minutes.

Given the proposed advantages to nonhuman females of mating with a range of male partners mentioned above, it might well be adaptive to possess motivational underpinning which would induce females to keep soliciting male partners. Long ago B. F. Skinner demonstrated that "intermittent reinforcement" produced remarkably stable behavior patterns, difficult to extinguish. Hence such a variable reward system as the female orgasm is not as maladaptive as it seems when we confine our imaginations exclusively to monogamous contests.

I see no basis for arguing that the female orgasm evolved to promote any breeding system in which humans are currently found. Certainly I would not argue that promiscuity is currently adaptive for women anywhere in the world,[43] but to say that the female orgasm is not currently adaptive among women does not mean that it *never* was adaptive for any of our prehominid ancestors. Nor does it preclude the possibility that an ancient adaptation might not have been secondarily incorporated into human breeding systems as an imperfect means to promote pair bonding. At the very least, our current state of ignorance concerning the comparative anatomy for female genitalia makes it premature to rule out the possibility that selection has operated in this area.[44]

Instead of the traditional anthropological approach of assuming that our ancestors resembled one or another extant primate (usually chimpanzees) and then attributing to our ancestors the attributes of the prototypes (for example, sexual swellings), and then speculating on what must have occurred in order to transform that creature into one with human attributes, I am proposing that we first examine the range of evidence currently available for all primates in an effort to locate general patterns. Although still probablistic rather than definitive, such reconstructions are less susceptible to personal bias than the arbitrary advocacy of a particular species as prehuman analogue.

As David Hamburg likes to remind students of evolution, "the most widely spread relic we have of early man is modern man."[45] When we examine our species today, we find a mildly dimorphic creature; men are between 5 percent and 12 percent larger on average than women. Men have relatively small testes and large penises. Women are receptive on a situation–dependent basis and are potentially capable of copulating on any day of the month. Receptivity is not signaled by conspicuous morphological signs.

At first glance such information would not appear to tell us much. However, comparative primatology suggests the following interpretations of these attributes. First, the common assumption that our ancestors were monogamous, living in obligate pair bonds cemented by the unique sexual attributes found among women is very likely wrong. The degree of sexual dimorphism in our species argues for the existence of mildly polygynous ancestors.[46] (However, based on the same reasoning, it would appear that there is a tendency towards decreased polygyny through time, since the degree of dimorphism among 4–million–year–old hominids like *Australopithecus afarensis* is apparently much greater than in contemporary humans.) Unless we are to plead for human beings as a special case, it seems unlikely that our ancestors lived in nuclear families with one male bonded for life with one female. Instead it appears that at least some males in the breeding population had the option of breeding with several mates.

Supposing we accept the proposition that our ancestors lived in either polygynous breeding units or in multimale breeding systems? The anatomy of the human testes renders the latter, multimale option less likely. Competition at the level of sperm does not appear to have been important in human evolution. Chimpanzees, with their large testes,[47] appear to have diverged from other apes in this respect; the most likely explanation is that larger testes were selected for as that species adopted a multimale, promiscuous mating system. By contrast, humans resemble the small–testicled orangutans, gorillas, and gibbons, suggesting that one–male breeding systems may have been the more ancient adaptation among hominoids.

Both species of chimpanzees, the pygmy chimp and the common chimp, differ from humans and all the other apes in exhibiting conspicuous swellings at midcycle. Like large testes, such swellings are probably an adaptation to a multimale breeding system. By a process of elimination then (see table 5), it is unlikely that our ancestors were characterized by either obligate monogamy or multimale breeding systems. By excluding alternative possibilities, one by one, the strongest likelihood is that our ancestors lived in either one–male polygynous breeding units or else they pursued some mixed strategy. This mixed strategy could have ranged from one–male "harems" with several breeding females, to perhaps singleton males living in social units containing no fertile females at all. This pattern would be compatible with the finding that contemporary women are characterized by an absence of discrete periods of sexual receptivity.

Contemporary attributes	Inferences
1. Relatively small testes and the absence of estrus	monogamous or unimale but not multifemale/multimale
2. Moderate sexual dimorphism	multifemale/multimale, or unimale, but not monogamous

Table 5. Contemporary human attributes, and inferences about the sort of breeding system in which they probably evolved.

POSTSCRIPT

I have attempted to place speculation about the sexual lives of early hominids into a broader framework, which takes into account our growing knowledge about other primates. Whenever possible, I have tried to balance my own views by mentioning alternative perspectives. Given the probablistic rather than definitive nature of knowledge in this area, such an open–ended approach is prudent.

But what does this effort to reconstruct the way of life of our ancestors have to tell us about our own lives? Some may find it fascinating, others distressing, that our ancestors 4 million years ago probably did not live in monogamous pair bonds that are highly valued in our own society. But even if correct, *this discovery has no bearing on contemporary decision making.* As has been pointed out many times, to say something evolved does not mean that it is desirable in the sense that it contributes to individual contentment or that evolved characters ensure survival of our species. One need only recollect that vast numbers of species are now extinct through utterly natural processes. Cereal advertising to the contrary, natural is not necessarily better. Furthermore—and again this is a point that should be obvious—hominids evolved under conditions radically different from those that pertain today.

A broader understanding of our biological legacy provides useful "background information" for those attempting to design their lives in line with rational or ethical guidelines. Such knowledge will yield more realistic expectation concerning present human conduct, as well as greater respect for the introspection and intellectual detachment that make civilization possible. But construction of the guidelines themselves remains outside the scope of a primatologist. If there is a moral to any of this, it could only be along the lines of Rosie's reply to her ne'er–do–well companion on *The African Queen:* "Nature, Mr. Allnutt, is what we are put in this world to rise above."

NOTES

1. This chapter is a summary of a long review titled, The absence of estrus in *Homo sapiens,* soon to be published by Smithsonian Institution Press (Washington, D.C.) in a volume titled, *The Origin of Humaness,* A. Brooks, ed. Those wishing fuller documentation of specific points might wish to consult this more detailed treatment.

2. Tutin, C. E. G. 1975. Sexual behavior and mating patterns in a community wild chimpanzees (*Pan troglodytes schweinfurthii*). Ph.D. Thesis, University of Edinburgh.

3. Symons, D. 1979. *The Evolution of Human Sexuality.* Cambridge University Press: Cambridge; Alexander, R. D., and K. M. Noonan, 1979. Concealment of ovulation, parental care and human social evolution. In: *Evolutionary Biology and Human Social Behavior.* N. A. Chagnon and W. Irons, ed.s Duxbury: North Scituate, MA; Strassmann, B. I. 1981. Sexual selection, parental care, and concealed ovulation humans. *Ethology and Sociobiology* 2:31–40; Turke, P. W. 1984. Effects of ovulatory concealment and synchrony on protohominid mating systems and parental roles. *Ethology and Sociobiology* 5:33–44.

4. Morris, D. 1967. *The Naked Ape.* Dell: New York; Lovejoy, O. 1981. The origins of man. *Science* 211:241–250.

5. Lovejoy, O. The origins of man. See note 4.

6. Morris, D. *The Naked Ape.* See note 4. Alexander and Noonan. concealment of ovulation, parental care and human social evolution. See note 3.

7. Alexander and Noonan. Concealment of ovulation, parental care and human social evolution. See note 3; Symons, D. *The Evolution of Human Sexuality.* See note 3. Benshoof, L., and R. Thornhill. 1979. The evolution of monogamy and loss of estrus in humans. *Journal of Social and Biological Structures* 2:95–106; Burley, N. 1979. The evolution of concealed ovulation. *The American Naturalist* 114:835–858; Hrdy, S. B. 1979. Infanticide among animals: A review, calssification, and examination of the implications for the reproductive strategies of females. *Ethology and Sociobiology* 1:13–40; Spuhler, J. 1979. Continuities and discontinuities in anthropoid–hominid behavioral evolution: Bipedal locomotion and sexual receptivity. In: *Evolutionary Biology and Human Social Behavior.* N. A. Chagnon and W. Irons, eds. Duxbury; North Scituate, MA.

8. Alexander and Noonan. Concealment of ovulation, parental care and human social evolution. See note 3.

9. See, for example, Strassmann, B. 1981. Sexual selection, parental care, and concealed ovulation in humans. *Ethology and Sociobiology* 2:31–40. Strassmann proposes what I call the "two–tiered–male, two–tiered–female hypothesis," a complicated extension of the Alexander and Noonan hypothesis. Strassmann assumed that our ancestors were polygynous at the time that concealed ovulation evolved. In her model, an elite strata of males has more success in mating than other males, and these males monopolize several mates each. Less successful, subordinate males–males of the second tier—were lucky if they mated at all. When they did mate, they channeled their efforts not into acquiring additional mates, but rather into caring for the few offspring they were able to produce. Some of the less desirable (second–tier) females, overlooked by dominant males because their sexual swellings were not sufficiently conspicuous, found themselves in the enviable position of producing offspring with fathers who helped, and consequently with offspring who had a better chance of surviving. As parental care became increasingly critical for the survival of the young, these second–tier, subordinate–but–helpful males became the most desirable; and it was the second–tier females with the least obvious signs of ovulation who were able to breed with them, since only such drab females could escape monopolization by dominant males.

10. Spuhler. Continuities and discontinuities in anthropoid–hominid behavioral evolution. See note 7.

11. See note 1.

12. Documentation for this reconstruction of australopithecines can be found in Johanson, D., and M. Edey. 1981. *Lucy: The Beginnings of Humankind*. Simon and Schuster: New York; Susman, R., J. Stern, and W. Jungers. 1984. Arboreality and bipedality in the Hadar hominids. *Folia Primatologica* 43:113–156; Walker, A. 1981. Diet and teeth: Dietary hypotheses and human evolution. *Philosophical Transactions of the Royal Society, London, B.* 292:57–64; Shipman, P. 1986. Scavenging or hunting in early hominids: theoretical frameworks and tests. *American Anthropologist* 88:27–43; Tague, R., and O. Lovejoy. 1985. The australopithecine obstetric pelvis. Paper presented at the 84th Annual Meeting of the American Anthropological Association, Washington, D.C., December 4–8, 1985; Trevathan, W. 1987. *Human Birth*. Aldine de Gruyter: New York.

13. Dahl, J. 1986. Cyclic perineal swelling during the intermenstrual intervals of captive female pygmy chimpanzees (*Pan paniscus*). *Journal of Human Evolution* 15:369–385.

14. Lemmon, W. B., and M. L. Allen. 1978. Continual sexual receptivity in the female chimpanzee (*Pan troglodytes*). *Folia Primatologica* 30:80–88.

15. Loy, J. 1975. The copulatory behavior of adult male patas monkeys (*Erythrocebus patas*). *Journal of Reproductive Fertility* 45:193–195; Hrdy, S. B. 1977. *The Langurs of Abu*. Harvard University Press: Cambridge, MA; Mori, A., and R. Dunbar. 1985. Changes in the reproductive condition of female gelada baboons following the takeover of one–male units. *Z. Tierpsychol* 67:215–224; Cords, M. 1984. Mating patterns and social structure in redtail monkeys *Cercopithecus ascaniuus*). *Z. Tierpsychol* 64:313–329.

16. Sekulic, R. 1982. Behavior and ranging patterns of a solitary female red howler (*Alouatta seniculus*). *Folia Primatologica* 38:217–232.

17. Andelman, S. 1987. Evolution of concealed ovulation in vervet monkeys (*Cercopithecus aethiops*). *The American Naturalist* 129:785–799; Van Noordwijk, M. 1985. Sexual behavior of Sumatran long–tailed macaques (*Macaca fasicularis*). *Z. tierpsychol* 70:277–296; Takahata, Y. 1980. The reproductive biology of a free–ranging troop of Japanese monkeys. *Primates* 21:303–329; Wolfe, L. 1984. Japanese macaque female sexual behavior; A comparison of Arashiyama East and West. In: *Female Primates: Studies by Female Primatologists*. M. Small, ed. Alan Liss: New York; Hearn, J. P. 1978. The endocrinology of reproduction in the common marmoset, *Callithrix jacchus*. In: *The Biology of Conservation of the Callitrichidae*. D. G. Kleiman, ed. Smithsonian Institution Press: Washington, D.C.; Brand, H. W., and R. D. Martin. 1983. The relationship between female urinary estrogen secretion and mating behavior in cotton–toped tamarins, *Saguinus oedipus oedipus*. *International Journal of Primatology* 4:279–290; Dixon, F. 1983. Observations of the evolution of behavioral significance of "sexual skin" in female primates. *Advances in the Study of Behavior* 13:63–106; Stribley, J. A. J. A. French, and B. J. Inglett. In press. Mating patterns in the golden lion tamarin (*Leontopithecus rosalia*): Continuous receptivity and concealed ovulation. *Folia Primatologica*.

18. Takahata. The reproductive biology of a free–ranging troop of Japanese monkeys. Figure 6. See note 17. Van Noordwijk. Sexual behavior of Sumatran long–tailed macaques (*Macaca fasicularis*). Figure 3. See note 17.

19. Schreiner–Engel, P., R. C. Schiavi. H. Smith, and S. White. 1981. Sexual arousal and the menstrual cycle. *Psychosomatic Medicine* 43:199–214.

20. This pattern was first reported by Udry, J. R., and N. Morris. 1968. Distribution of coitus in the menstrual cycle. *Nature* 220:593–596; Adams, D. B., A. R. Gold, and A. D. Burt. 1978. rise in female–initiated sexual activity at

ovulation and its suppression by oral contraceptives. *New England Journal of Medicine* 299:1145–1150; Wortham, C. 1978. Psychoendocrine study of human behavior: some interactions of steroid hormones with affect on behavior in the !Kung San. Ph.D. Thesis, Harvard University; Matteo, S. and E. F. Rissman. 1984. Increased sexual activity during the midcycle portion of the human menstrual cycle. *Hormones and Behavior* 18:249–255.

21. Michael. R. P. 1972. Determinants of primate reproduction. *Acta Endocronol. Copenh.* Supplement 166:322-363. See fig. 2.

22. Even here, the distinction is somewhat blurred by the fact that seasonal peaks in births are reported for some human populations; for example, peaks in conception occurred during late spring and early summer throughout Europe between the sixteenth and nineteenth centuries. See Wrigley, E. A., and R. S. Schofield. 1981. *The Population History of England, 1541–1981.* Harvard University Press: Cambridge, MA 286–305.

23. The postpartum cessation in sexual activity is both longer and more pronounced in nonhuman primates, even though a pattern indicating some reduction in mating is universal to both human and nonhuman primates. For example, recent studies of Western women who breast feed have shown that the mean rate of intercourse drops from an average of 2.5 to 1.2 times per week in the months after birth. In some non–Western societies, postpartum taboos on sexual behavior are the rule. See Stern, J. M., and S. R. Lieblum. 1986. Postpartum sexual behavior of American women as a function of the absence of frequency of breast feeding: A preliminary communication. In: *Proceedings of the Tenth Congress of the International Primatological Society.* J. Else and P. Lee, eds. Cambridge University Press: Cambridge, England. 3:315–324. Alder, E., and I. Bancroft. 1983. Sexual behavior of lactating women, a preliminary communication. *Journal of Reproduction and Infant Psychology* 1:47–52.

24. Anderson, C. 1986. Female age: male preference and reproductive success in primates. *International Journal of Primatology* 3:305–326; Van Noordwijk, M., and C. P. van Schaik. 1986. Comparisons of primate sexual behavior: The primate roots of human sexuality? Paper presented at the American Association for the Advancement of Science, Philadelphia; Scott, L. 1984. Reproductive behavior of adolescent female baboons (*Papio anubis*) in Kenya. In: *Female Primates: Studies by women primatologists.* M. F. Small, ed. Alan Liss: New York.

25. Dixson, A. F. Observations of the evolution of behavioral significance of "sexual skin" in female primates. See note 17. Wright, P. C., M. K. Izard, and E. L. Simons. 1987. Reproductive cycles in *Tarsius bancanus. American Journal of Primatology* 11:207–215.

26. Clutton–Brock. T., and P. H. Harvey. 1976. Evolutionary rules and primate societies. In: *Growing Points in Ethology.* P. P. G. Bateson and R A. Hinde, eds. Cambridge University Press: Cambridge, England.

27. Ibid.; Hrdy, S B. 1979. Infanticide among animals: A review. classification, and examination of the implications for the reproductive strategies of females. *Ethology and Sociobiology* 1:13–40.

28. Darwin, C. 1871. *The Descent of Man and Selection in Relation to Sex.* Murray: London, England.

29. Smith, R., (ed.). 1984. *Sperm Competition and the Evolution of Animal Mating Systems.* Academic Press: New York; Halliday, T., and S. J. Arnold. 1987. Multiple mating by females: A perspective from quantitative genetics. *Animal Behavior* 35:939–941; Hrdy, S. B. 1981. *The Woman That Never Evolved.* Cambridge University Press: Cambridge, England; Small, M. In press. Is there

really sperm to spare? *Current Anthropology.*

30. See, Ransom, T. W., and B. S. Ransom. 1971. Adult male–infant relations among baboons (*Papio anubis*). *Folia Primatologica* 16:179–95, and especially the exceedingly rich and well–documented reports in Altman, J. 1980 *Baboon Mothers and Infants.* Harvard University Press: Cambridge, MA; and Smuts, B. 1986. *Sex and Friendship Among Baboons.* Academic Press: New York.

31. Busse, C., and W. J. Hamilton, III. 1981. Infant carrying by male chacma baboons. *Science 212:1281–1283.*

32. Taub, D. (ed.) 1984. Primate Paternalism. Van Nostrand: New York; Terborgh, J., and A. W. Goldizen. 1985. On the mating system of the cooperatively breeding saddle–backed tamarin (*Saguinus fuscicollis*). *Behavioral Ecology and Sociobiology* 16:293–299.

33. Hrdy, S. B. 1979. Infanticide among animals: A review, classification, and examination of the implications for the reproductive strategies of females. See note 27. Leland, T., T. Struhsaker, and T. Butynski. 1984. Infanticide by adult males in three primate species of the Kibale Forest, Uganda: A test case of hypotheses. In: *Infanticide: Comparative and Evolutionary perspectives.* G. Hausfater and S. B. Hrdy, eds. Aldine de Gruyter: New York.

34. Short, R. 1979. Sexual selection and its component parts, somatic and genital selection, as illustrated by man and the great apes. IN: *Advances in the Study of Behavior.* J. S. Rosenblatt. R. A. Hinde, C. Beer, and M. C. Busnel, eds. Volume 9, Academic Press, New York; Harcourt, A. H., P. H. Harvey. S. G. Larson, and R. V. Short. 1981. Testis weight, body weight, and breeding systems in primates. *Nature* 293:55–57.

35. Terborgh and Goldizen. On the mating system of the cooperatively breeding saddle–backed tamarin (*Saguinus fuscicollis*). See note 32.

36. Elsewhere I review species outside of the primate order (several rodents and one species of louse) in which receptivity is also situation–dependent: see note 1.

37. This model was first proposed by Desmond Morris and later updated and refined by Owen Lovejoy, and can be found in most major textbooks in physical anthropology. As an example, see Nelson, H., and R. Jurmain. 1985. *Introduction to Physical Anthropology.* West Publishing Co.: Saint Paul, MN.

38. Mascia–Lees, F., J. H. Relethford, and T. Sorger. 1986. Evolutionary perspectives on permanent breast enlargement in human females. *American Anthropologist* 88:423–428.

39. Galdikas, B. 1979. Orangutan adaptation. In: *The Great Apes.* D. Hamburg and E. McGown, eds. Benjamin Cummings: Menlo Park, CA; De Waal, F. B. 1987. Tension regulation and nonreproductive functions of sex in captive bonobos (*Pan paniscus*). *National Geographic Research* 3:318–335.

40. See Chevalier–Skolnikoff, S. 1974. Male–female, female–female, and male––male sexual behavior in the stumptail monkey, with special attention to the female orgasm. *Archives of Sexual Behavior* 3:95–116; for stumptailed macques see Burton, F. 1971. Sexual climax in female *Macaca mulatta.* In: *Proceedings of the Third International Congress of Primatology, Zurich, 1970.* S. Karger: Basel, Switzerland. 3:18–191; Michael, R. 1972. Determinants of primate reproduction. *Acta Endocrinologia,* Copenhagen Supplement 166:322-363; Allen, M. L., and W. B. Lemmon. 1981. Orgasm in female primates. *American Journal of Primatology* 1:15–34.

41. See Masters, W., and V. Johnson, 1966. *Human Sexual Response.* Little, Brown and Co.: Boston, MA: Hite, S. 1976. *The Hite Report.* Macmillan: New York, for data showing that relatively few women reach orgasm from intercourse

alone. For the quote on female satisfaction, see Pugh, G. 1977. *The Biological Origin of Human Values.* Basic Books: New York. 248.

42. The more important controversy for biologists does not center on whether or not the female orgasm evolved to cement pair bonds (clearly, it did not), but on whether it evolved at all. See, for example, Symons, D. *The Evolution of Human Sexuality,* see note 3; Hrdy, S. B., *The Woman That Never Evolved;* see note 29. *Gould, S. J. 1987.* Freudian slip. *Natural History.* (April): 14–21; Alcock, J. 1981. Ardent adaptationism. *Natural History.* (April): 4. Although there continues to be debate on the point, I accept as given that orgasm does occur in nonhuman females (see note 40). Nevertheless the occurrence of orgasm among females would appear to be, if anything, even more erratic and inefficient in other animals than it is among humans. Only some nonhuman females occasionally experience orgasm, and when they do, it requires greater stimulation than is typical of a single mating bout in the wild. This variable and inefficient response leads some (among them biologist Steve Gould and anthropologist Donald Symons) to conclude that orgasms are nonadaptive, a potential present in females because it is selected for in males. Based on the same evidence, others have argued that the female orgasm could well be adaptive (for example, biologist John Alcock). Others (including myself) have suggested that the female orgasm is no longer adaptive, but may be a vestige of a response that was adaptive in a prehominid condition when our ancestors lived in breeding systems comparable to those of the monkeys and apes I describe here.

Discussion in this lively debate centers about the clitoris, a unique organ whose only known function is to translate sensual stimuli into physical and psychological sensations. The "nonadaptive" camp stress the fact that developmentally, the clitoris is nothing more than a homologue of the male glans penis. Like nipples present on the chests of men, this organ is incidentally present in one sex because it is essential to the reproductive system of the other.

In contrast, the "once–was–adaptive" camp notes that if the clitoris is nothing more than a homologue of the penis, we might reasonably expect that across species the incidental organ (the clitoris) should track evolutionary changes in the critical one (the penis). As it happens, evidence from a variety of species is consistent with this notion; the clitoris does indeed exhibit various idiosyncratic traits (such as miniscule spines) that are also characteristic of males belonging to the same species. Such evidence does indeed support the notion that the two organs are homologues. But consider the apes. In the gibbon, orangutan, gorilla, and chimpanzee, the clitoris is generally well–developed, both absolutely and relatively larger in these anthropoid apes than in women (See McFarland, L. G. 1976. Comparative anatomy of the clitoris. In: *The Clitoris.* T. P. Lowry, and T. S. Lowry, eds. Warren H. Green: Saint Louis, MO.) The clitoris and labia minora (homologous to the urethral surface of the penis) are especially well–developed in the promiscuously mating chimpanzee. Yet in males of this same species, the penis is strikingly narrow compared to its relative and absolutely much greater size in the related species, *Homo sapiens.* Penis and clitoris, then, are not particularly homologous in chimpanzees, and their differences raise the distinct possibility that selection has indeed acted independently on male and female organs in the species. Recently Leslie Digby of the University of California, Davis, has called to my attention similar cases among monogamous and polyandrous species of New World monkeys, in which males and females exhibit quite different genital morphologies. These include *Aotus, Callimico, Leontopithecus,* and *Saguinus.*

43. Note that I do not consider prostitution, which I consider to be an economic survival tactic rather than a reproductive strategy.
44. See note 42, above.
45. Hamburg, D. 1961. The relevance of recent evolutionary changes to human stress biology. In: *Social Life of Early Man.* S. Washburn, ed. Viking Fund Publications in Anthropology 31:278–288.
46. Harcourt, et al. Testis weight, body weight, and breeding systems in primates. See note 34.
47. Ibid.

PANEL DISCUSSION

MARGULIS: You began your address by discussing pheromones in nonhuman mammals. I am wondering what is the status of research on sexual chemical communication in primates?

HRDY: There has been very little work done on pheromones among wild primates. Goldfoot has found that in captive macaques, chemical communication appears to be important, but not essential for reproduction. He fixed his animals so that they could not smell, and then he went on to see if they would breed on the basis of visual clues alone. They do. Humans have chemical scents very similar to those of macaques, and they too change with the menstrual cycle, but I suspect that most people do not notice them. I would argue that like macaques, our scents are not an essential part of our sexual lives, but we clearly do not know enough about them.

AUDIENCE QUESTION: Is sexual selection acting on present–day humans?

HRDY: Surely natural selection is at work on present–day females and it is weighing heavily on women like myself, who have a career outside the home. We are producing fewer children than those women who are producing five, six, twelve kids. Natural selection still operates.

Sexual selection in the sense of male–male competition may or may not be operating. If you look at particular cases—like Hunt, the famous oil millionaire, who had several wives because he wanted to pass on his genius genes—then you would say yes. Clearly that observation has to be discounted by all of the millionaires who do not produce lots of children. I have seen several statistical analyses of this question and they produce conflicting results. There is, however, enough evidence to suggest that sexual selection may be operating on males.

Is sexual selection operating on females? It is hard for me to answer that, because I am so committed to the notion that variance in female reproductive success is presently being underestimated by evolutionary biologists. I guess I would have to say that I do not know.

HAMILTON: It always has seemed to me that in human beings

where property and status are passed down to a large extent, that there is an unusual opportunity for sexual selection to be operating in females almost to the same extent as in males. If you look at the first generation down, it is true that the woman is limited by her physiology—that is, how many babies she can produce. If you look two or three generations down, a female may be as different from a male in the number of descendants as some other extremely successful male.

HRDY: I agree completely that theoretically that may be the case. My point is that we do not have the evidence.

HAMILTON: Is that because nobody has looked for it, or have people tried?

HRDY: They have not looked carefully enough.

HAMILTON: Well, they should, because it seems to me that the signs are that sexual selection has been uniquely powerful in the human female. It had produced the kinds of characters one would only expect in a competitive male, yet the human male does not have those same characters.

AUDIENCE QUESTION: How prominent is rape in mammals?

HRDY: The famous cases of forced copulations are from mallard ducks, and I know very little about birds. In nonhuman primates, the only well–documented cases are from orangutans. At midcycle, a female orangutan will seek out the dominant male in her vicinity. At other times during her cycle, young males and subordinate males will sometimes cluster around her and it appears that she is being forced to mate with them. She resists, she screams, she fights, but they mate with her anyway. It is either a good act, or it's the equivalent to rape. Primatologists try not to use that term. People tend to get upset.

AUDIENCE QUESTION: How common is homosexual behavior in primates?

HRDY: There is a great deal of homosexual behavior in the natural world, both male–male and female–female. What we have yet to document is a lifetime strategy of homosexual behavior. It is usually associated with a heterosexual strategy. We simply do not have long–term lifetime information for individual monkeys that would allow us to say that in any primate species you have exclusive homosexuality.

AUDIENCE QUESTION: Could you elaborate on your explanation for the evolution of the female orgasm?

HRDY: Recently Stephen J. Gould wrote a piece in *Natural History* magazine called "Freudian Slip." Although I did not agree with his analysis, he did us all a favor by bringing out into the open something that has been a taboo subject. Orgasm is something that needs to be studied; but if any graduate student went out to study orgasm, there would be such a stigma attached to their work that they probably would not be able to get a job. So I am grateful to Dr. Gould for bringing it out into the open.

Even so, I disagreed with Dr. Gould's conclusions. I had suggested that the female orgasm might have evolved in the context of the sexual strategies I have discussed in my address. I feel that orgasm is part of the emotional underpinnings necessary to keep a female soliciting a number of different male partners. Gould argued that the orgasm is nonadaptive, that human females have a clitoris because males have a penis. In other words, the clitoris is a homologue of the penis. I would suggest caution at this very early stage of our understanding. While there is abundant evidence that the two structures have evolved together, when you compare across species you find—for example, in the case of humans and chimps—a large penis and small clitoris in humans, but a small penis and large clitoris in chimps. If these organs are truly homologues of one another how do you explain such cases? There must have been independent selection on the clitoris.

QUESTIONS FOR FURTHER DISCUSSION

1. Given the history of bias in the design and reporting of primatological studies, is there any way to make progress toward understanding primate sexual evolution?
2. Much of the sexual politics of nonhuman primates is based on behavior many humans would consider immoral (polygyny, polyandry, infanticide, and so forth). How does our distaste (or maybe "thirst") for such stories influence our understanding of the natural world?
3. What are the pitfalls of using the comparative approach employed by Hrdy to reconstruct the past? What are the advantages?

4. Would you use the information Hrdy provides to design human social interactions (for example, to draft legislation or design housing complexes), or is this information irrelevant to human concerns?

5. In the panel discussion, Hamilton suggests that sexual selection has had a strong influence on the human female. What does he mean by this, and what female characters does he have in mind?

6. Why is there so much uncertainty as to the origin of human traits?

Philip Hefner takes a somewhat different approach to the topic of the evolution of sex. In this chapter, Hefner describes the role of recombination of genetic material in the greater scheme of universal thermodynamics. He speaks not as an evolutionary biologist, but as one who has dealt with the issues of the role of God in the evolutionary process.

As a nonbiologist, Hefner uses biological concepts in a slightly different context. Recognize that Hefner is speaking of a level of analysis far above that available to those who study the creatures of the earth. His emphasis is on the interpretation of the role of sex rather than the reasons for the success of sex in the sexual organisms with which we are familiar.

The Editors

6. Sex, for God's Sake: Theological Perspectives

PHILIP J. HEFNER

Shortly after Christmas, I asked an acquaintenance of mine—a professor of genetics at a leading American university who has herself published on the subject of sex—to help me find a way on this question of the evolution of sex. In a letter, she reviewed the positions of two of the other contributors to this volume and concluded, "Nobody has any claim to fundamentality on this topic, to my mind: we're all more or less speculating. The problem with most that's been written, I suspect, will be its relative opacity to the nonscientist." Some weeks later, after I had digested for myself a number of books and articles on the theme, including several very impressive pieces by members of this 1987 Nobel symposium, I called a friend who is a biologist at the University of Chicago. I laid out for her the learnings from my reading that seemed important to my own presentation, only to have her respond, "Phil, I wouldn't use those ideas as a foundation, because the mathematics that they're based on hasn't been worked out carefully enough yet."

I review these incidents in order to make three points. First, about the topic that has called us together, "The Evolution of Sex." Clearly, it is an important issue, and utterly fascinating. But more than this, it is one of genuine excitement, because although it encompasses one of the most common dimensions of life—human and nonhuman—it defies final explication by even the finest scientific investigators. If ever the misused term "cutting edge" were appropriate, it would be for the issues we are discussing here.

A second consideration from the incidents I recounted throws light on the kind of enterprise in which we are engaged. The members of this discussion are all laypeople to various degrees, all seeking to understand the issues in an area where highly specialized expertise is a requirement, but also an area where that expertise can only be communicated with, as the geneticist wrote, "relative opacity" to the layperson. Everyone is some sort of layperson. Some of us lack expertise in all of the sciences represented here: others are bona fide scientists in one or more of the relevant fields, but not in all. Others are relatively uninformed about religion and even less well informed about theology, ethics, history, or politics, as they pertain to our issues.

Our lack of scientific understanding may be the greatest problem. People like myself are ones who try to be alert to the basic issues of life today, who care about the issues, and who in many cases are in positions to make decisions that touch on the issues, all the while remaining in the shadowy precincts of nonexpert. Most of our elected officials and schoolteachers are in the same situation. This dilemmma is compounded when we are discussing matters on which, as my two scientific consultants indicate, even the scientists are not of one mind, in which "all are more or less speculating." This is, however, no excuse for the laypeople to sit back in indifference. That the scientists themselves lack consensus on some of the most important issues pertaining to the evolution and meaning of sex does not lessen for a moment the importance of the theme for all of us! We all need to know more, and sometimes even the most esoteric scientific discussion presents issues we have to act upon as laypeople. The situation simply means that we all are explorers, together but at different levels, on the same pathway. This situation characterizes most of our attempts to be responsible in our advanced technological civilization, which now dominates all of planet Earth.

For the conversation between theology and science, however (and this is the third point I wish to draw from my efforts to understand the evolution of sex), this means particularly that theological interpretations, although they must be made, must be tentative. Theology reflects upon the facts of life. It does not fabricate those facts. When the Hebrew prophet Isaiah was certain that Cyrus, as conqueror of Babylon, would recognize the Hebrew God, Yahweh, and release the exiles, he prophesized that there would be a triumphant procession of exiles, on a veritable highway across the desert, returning to restore Jerusalem: the high places would be made low, and the rough places would be smoothed out for the returning pilgrims. Less than two years later, when it became clear that Cyrus was politically sensitive enough to favor the god Marduk, whom his subject people had been worshiping in large numbers, and that he would permit only a trickle of exiles to return (with neither large subsidies nor limousines across the desert), the prophet changed his message. He produced, rather, the striking message of Israel as a Suffering Servant, who would know triumph only through vicarious sacrifice for the other nations. The facts of life required the prophetic reassessment of the preacher's own message about Yahweh. We honor both of his attempts in the church, reading them at different high festivals, the one in Advent, the other in Holy Week. We must be clear, however, about the relative opacity of the events that called forth his theology and revision of even his most beautiful utterances.

So it is with theology and the evolution of sex. Interpretations must be given, and they will be given. However, theological interpretations must acknowledge the state of our understandings of the facts of life. As the scientists knows a thing or two about having hypotheses falsified and revised, so must the theologian be prepared to enrich their understanding as we get clearer understanding of the facts of life. Just as it is no insult to nature when scientists must rearrange their concepts, so it is no reflection on God when theologians do the same. For the scientist, the rearranging activity is a mark of faithfulness to the data of nature; for the theologian, the same should be said: it is a mark of faithfulness to what God has unfolded for us about the world in which we live and about ourselves through the knowledge that scientists are able to grasp.

The relevance of this dynamic, ambiguous, and tentative character of our knowledge about sex for this theological presentation may be elaborated further if we ask the question, "Just what is a theologian doing discussing this theme?" I suppose that there are a variety of expectations on this score. St. Thomas Aquinas's description of theology seems most useful in this setting: theology is that enterprise that relates all things to God. Theology is distinguished not by what things it talks about, but rather by the perspective from which it talks about them. It focuses on things that others also find important—for example, politics, human life, good, evil, and sex. What makes discourse theological is that it views those things in terms of their relationship to God.[1]

Now just what in our theme is to be related to God? The answers given even by members of this symposium are varied. Sexual recombination of genes is the subject under discussion for some of our scientists, although there is not agreement on whether this recombination is a successful adaptive strategy or whether it is a legacy we bear within us that emerged in situations that had no bearing on sex as such, but to which sex is a response. These scientists are clear that the referent for the word "sex" is not the act of reproduction or gender or events of pleasure, rather it is *mixis*, mingling. Michael Ghiselin calls it "a reorganization of the genome",[2] Lynn Margulis describes sex as "a process characteristic of live organisms only; the complex set of phenomena that produces a genetically new individual, an individual that contains genes from more than a single source."[3] Sarah Blaffer Hrdy certainly does include reproduction and the elements of gender, physiology, psychology, and behavior related thereto in her definition of sex. She does introduce the theme we might call "sexuality." Sex as an item of culture and society has gotten some attention. Very little has been even hinted about sex as having to do with love. The contributions to this symposium volume have tended to move progressively from micro to macro considerations. I suppose that the theologian is considered to be at the top of the macroreality!

The function or purpose of sex cannot be separated from a discussion of what the theologian can do to relate sex to God, because what a thing *is* cannot be divorced from what it *does* and what it *is for.* If we set out to understand the automobile, for example, we would be surprised if discussion focused entirely on its shape and structure and how it works, without ever mentioning that is a means of transportation. I gather from my reading and listening to scientific discussions that scientists are not nearly as clear on what sex is for as they are on its mechanisms and structures. Or perhaps we should say that scientists give multiple answers to the question of sex's purpose.

Official theologians of the Roman Catholic church would supply their own concepts of what sex is for, namely for reproduction and the enhancement of love and faithfulness within the marriage bonds. Furthermore, they would argue that this is what is really important about sex. I shall not take up this sort of argument, although I do believe that it would be a responsible way of proceeding. After all, regardless of what the "harder" sciences say about the evolution of sex, most human beings engage in sexual behaviors because of their pleasures (whether inside the marriage relationship or not), and some elements of sexual behavior are very nearly the only strategy the human species has for reproducing itself. These traditional concerns are not central to the scientific issues being raised by the contributors to this volume.

Given the great importance of sex, the theologian has no choice but to supply suggestions as to what sex is for, even if the scientists are less than clear about this themselves, and hope that these suggestions are not fundamentally in opposition to what the scientists tell us. I will speak of these purposes of sex, together with what the scientists tell us about its mechanisms and structures, in terms of their possible relation to God.

SEX PLACED WITHIN A THERMODYNAMIC FRAMEWORK

I shall follow the strategy of supplying a suggestion concerning the purpose of sex, but I shall follow rather closely what I perceive to be the prevailing view of the scientists, in that I shall speak of the purpose of sex as the recombination of the genome. Ursula Goodenough speaks of *transformation* as a possible synonym for "recombination."[4] Let us for the moment consider that the purpose of sex is *the transformation of genetic composition through the births of offspring that result from the mixing of the genetic material of their parents (thus the offspring differ genetically from their parents) rather than through simple genetic replication by one parent.* What I want to say is that this sex can be understood within the purposes of God. That is why I speak of "sex, for God's sake," that is, sex for the purposes of God. The way I shall unfold my argument may be rather tortured, however, and I beg you to bear with me.

It seems to me that we make progress in understanding the phenomenon of sex if we place it within an even larger evolutionary context. One scientific thinker who has recently done just this is Jeffrey S. Wicken, in his new book, *Evolution, Thermodynamics, and Information: Extending the Darwinian Program.*[5] The larger evolutionary context is that of the entire physical world. Wicken brings to bear thermodynamics (the science of process and stability) and information theory (the science of structure and complexity). What results from such an approach is a contextualized picture of evolution, including the familiar elements of variation, selection, and adaptation that we associate with the processes of evolution. The contextualized picture includes *thermodynamic flow which pertains to patterns of energy transformation* (which accounts for the macroscopic shape of evolution); and *individual behaviors* (which accounts for the microscopic shape of evolution). "Ecosystem flow patterns cannot develop . . . except through individual or group strategies for surviving and leaving offspring. Conversely, the competitive success of ecosystemic flow patterns imposes selective conditions on the evolution of individual adaptive strategies."[6]

When sex as genetic recombination through more than one parent is viewed in this larger context, some very interesting observations may be made. Sexual recombination of the genome is an instance of the "entropic randomization" which is a "driving force of evolution."[7]

In this sense as *mixis*, the connection of sex to the entropy principle is so fundamental that one could argue that this phenomenon is not only primitive to life, but that it derives from a universal law that only particularized itself in living systems. Indeed, in the tendency of abiotic microspheres to fuse and exchange materials we see protosex merge with the flow nature. Given the primordial nature of mixis–randomization, explanations for the evolution of sexual reproduction needn't begin with nothing. They can begin with the primitive tendency for genomes to randomize under the entropic drive to access new configurational microstates: again, the unity of cause and diversity of context that characterizes evolution.[8]

Wicken carries the argument further with the assertion that sex is "a condition for the exploration of the biosphere."[9] A little explanation may be helpful here. Physics, as I learned it almost forty years ago in high school, spoke of entropy as the running down and depletion of energy in closed systems. Even today, that is what many people outside the scientific community think of when they hear of entropy or the Second Law of Thermodynamics. There is much more to this difficult concept, however. One of the most important ways in which entropy can be understood is the process of emerging alternations that make for possibility. Physicists may define entropy as the increase of possibilities for transformation or organization, or for the increase in the number of "accessible states."[10] As such, entropy seems to be a condition for life as we know it. For example, it is the coexistence of hot and cold in an unstable, nonuniform, and noncontracting universe that enables the heat transfers and the radiational effects that make life possible.

This drive for randomization is a drive "to access new configurational microstates."[11] There are, however, constraints on this drive for randomization, and they are provided by the structures in which recombination is carried out, as well as by selection and adaptation. Thermodynamic forces for randomization form the macroevolutionary process in which sex takes place; whereas the constraints of specific genetic structures and genetic processes, and processes of selection and adaptation with groups and individuals, form the microevolutionary process. The macro and micro interact. Individuals and groups, on the one hand, are caught up within the thermodynamic flows of the larger systems, and those flows impose conditions upon them; on the other hand, that flow cannot itself develop except through the constraints that are constituted by groups and individuals.

Such a perspective may organize some of the observations about sex and how its benefits and disadvantages are related to both groups and individuals. Placing sex within the larger evolutionary framework may account for how the costs and benefits of sexual recombination are distributed within the hierarchical context, how the payoffs seem to be more clearly visible at the macroevolutionary level than at the micro level. Natural organizations, "from organisms to societies" find that their "existences are inseparable from their operations as informed dissipative structures."[12]

Individuals operate within an operational hierarchy in which selfish interests are located within a broad framework of functionality ultimately connected to the flow of energy and resources through the biosphere. They also operate within a *historical* context that provides for the *taxonomic* hierarchy ... Sex is the bond connecting individual and species that has made the taxonomic hierarchy possible; the part–whole relationship it mediates provides the basis for species selection.[13]

There is more going on in sexual behavior and within structures of sex than the strategies of individuals. There is also the process of the interaction of larger systems and the largest physical reality of all, thermodynamic flow. Analogously, we might say that in earthquakes, there is more going on than teacups rattling and walls falling. Masses of the earth interact along fault lines. Beyond this, still larger geological systems interact as they become the locus of thermodynamic flow.

It would be difficult for a teacup to sense this, or even for the owner of a favorite teacup. Conversely, the larger earth masses are more interested in moving mountains than teacups; but if teacups are all that are available for moving, the larger earth masses must settle for that. Teacups are the constraints, the large earth masses are the macroprocesses.

When we put matters in this perspective, we are understanding the process of evolution hierarchically, in the sense that its processes are constituted by the dynamics of several systems all at once, from the microscopic to the macroscopic. In the recombination or transformation of genetic systems, "genes are propagated in the context of interlacing flow patterns," and we do not properly understand this except through a "perspective of contextualized functionalism."[14]

Sexual recombination makes the possibilities of the entire population available for the individual through its entropic dynamics. Sexual recombination makes it possible for evolution in the biosphere to be progressive and "adaptively versatile."[15] These payoffs are at the macroevolutionary level, possible because of the thermodynamic flow within which all organisms and societies live and move and have their being. Wicken writes, "The warrant for sex is its *creative power*, its contribution to the kinds of genomic reorganizations that moved fish to land and reptiles to air—creating the very ecological space in which life could spread. It is the stuff of history."[16]

THE THEOLOGICAL TURN

Why would we wish to further complicate the already complicated ideas of the evolution of sex by placing them within the conceptual framework of the larger system of physical reality and its thermodynamic properties? In response, we must acknowledge that micro–level interpretations of sex simply are not enough to clarify some of the thorniest problems for us. Larger systems are at work in the processes of sexual recombination. Of all people, the theologian is certainly aware of this. Theology cannot leave mircophenomena uninterpreted by macroscopic conceptualities.

It is not possible for me as a theologian to make a theological interpretation of the microprocesses described by the other contributors to this volume unless I understand more clearly the larger significance of those processes. Introduction of thermodynamic concepts helps me at this point.

It is important to understand that the bacterial studies, the population biology, the botany, the primatology, and all other scientific disciplines are also entities that transpire within larger movements of reality. Do not misunderstand me—the existence of larger wholes does not get us to God. I am not trying to present any proofs for God in this argument. However, placing sex within the thermodynamic scenario does provoke us to think big, and thinking big at least puts us in proximity to what the theology talks about. Such a scenario also helps us to see that it is not unusual that we should bear within us a multilayered legacy, as the scientists here have suggested.

Whether explicitly or implicitly, theology operates with a concept of "human being," which influences all that it has to say about such issues as sex. Our understanding of human being flows from what we have already considered. *Homo sapiens* appears in the evolutionary scheme of things as a two–natured creature, the confluence of two streams, the genetic and the cultural. This confluence has formed a powerful, if tense alliance within the central nervous system and the brain of the human person. Some have spoken of a symbiosis of genes and cultures as the most important characteristic of human being.[17] Humans are still trying to understand this interconnection of genes and culture, and are not yet clear on how the two can coexist in a wholesome and life–promoting manner.[18] What does seem clear is that human culture is always seeking to stretch or transform the physical and biological components so as to maintain their survival, while at the same time to enable them to become something new.[19] This stretching represents the process by which the physico–biological spheres will either be distorted, destroyed, or enhanced through human culture's opening up for them a destiny that is their fulfillment.

The human cultural dimension of evolution also operates within the process of thermodynamic flow, the locus of tremendous forces of entropic randomization. This powerful dislocation, or "mixed–upness," is an expression of the entropic drive to increase the number of accessible states, actualize future states that will enable the most desirable existence for the whole complex of macro, meso, and microsystems. The concrete shapes of historical existence in which humans find themselves constitute the constraints with which the macro–flow must come to terms.

Much more needs to be said about culture, and indeed this would be a suitable topic for another Nobel conference. This present conference is itself, however, an act of cultural self–transcendence, an act of the human animal pondering what it means to be a sexual creature within the thermodynamic flow.

Christian theology does indeed relate this process to the creative and redemptive work of God. In the Christian purview, the entire thermodynamic order came into being by God's free agency, as we express this in the doctrine of Creation Out of Nothing. This doctrine affirms that the created order is finally dependent for its entire existence upon God. The ongoing presence of God as the source upon which all existence is dependent is declared in the doctrine of Continuing Creation. Christian faith wants to affirm that the dynamics and structures of the thermodynamic flow, of the physico–chemical processes, of biological evolution, and of cultural development are the actual instrumentalities through which God continues to be creative. God is glorified through the tradition as the One who can bring new things into being. The explosions of chaos or mixed–upness that mark entropic randomization, even in their verbal imagery, are not foreign to what we affirm of God's persisting work of transformation. On the other hand, Christian tradition also knows the God who is faithful to concrete structures of history, even as divine power is facturing them so that the new can be born. The faithfulness of God is the ground of our trust. Our trust is that the newnesses of God are genuinely fulfillments of what we have been and are, even when the novelty feels like death.

Again, Christian faith knows full well that the new comes into being through death: it does not consider that death to be failure or defeat, but rather to be the pathway of recreation and new creation.[20] The defective state of human efforts is given serious regard in the various doctrines of sin, which point to human finitude, weakness, misguidedness, and downright perverseness. This defect is the ambience within which death appears. No more than death, however, is this defect necessarily the ground of defeat, since the God of the process is merciful, as well as faithful. Nevertheless, the defect can be the ground of human extinction, since the human creature is free and not coerced, insofar as culture is concerned. Human culture can, through ignorance or malevolence, destroy its biotic cosymbiont and the environment in which they live. Christian faith knows full well that this doleful end to *Homo sapiens* is a possibility, even as it believes that such an end is not necessary.

The concept of human being that Christian faith holds might well be called that of the *created cocreator*. The human being is the product of a determined system, whether that system be conceived as God's agency, the evolutionary nexus, or both. This human creature enters into the further movement of the evolutionary trajectory, however, in a way that no other creature does—through the agency of culture. We call this cocreatorhood, because human culture opens up genuinely new possibilities for existence for the natural world. Whether it is opening the possibility of coal tar becoming the fabric of panty hose and polyester suits, of iron ore and other minerals becoming guns and ICBMs, or of turning natural compounds through chemistry into medicines that can heal the body and the spirit, this activity of human culture is the most powerful determinant of our existence today. This cocreatorhood is as fully willed by God as the rest of creation. It is provocative to think that God should bring it about, that the created order is not enslaved forever to programmed instructions, but that in the human species a zone of freedom should emerge which enables the creation to share in its own further creation and freely choose to honor or to repudiate its Creator. The stretching and possibly fulfilling activity of the human being through its culture is the most striking expression to date on planet Earth of God's calling forth a cocreator.

The Christian faith knows that the existence of this created order is fundamentally shaped as much by what it can become as by what it had been. Theology calls this the eschatological dimension of reality. Nature is not content with what has been, and human culture even less so. Faith holds that God's eschatology will be the transformation and fulfillment of what has been. The most fundamental norm for human culture, in the eyes of faith, is whether its stretching and bending of the physico–biological reality is consistent with the fulfillment that God wishes to work. Though he lived in a much simple time, when the pressures on the cocreator were not so intense, we believe that Jesus of Nazareth is the prototype of what the cocreator is to be. That is why we call him the Christ, the Second Adam, who reveals what all of us Adams and Eves are meant to be.

SEX AS A PARADIGMATIC REALITY

I hope that the rather dense theological discussion that I have just concluded makes clear that I have spoken of God and of the human being within God's creative aims in ways that are of a piece with what we already suggested about sex. Sex is itself a microscopic instance of the larger processes that I have identified as characterizing the created cocreator and also God. Sexual recombination or transformation of the genetic realities is itself a stretching of physico–biological stuff. The created cocreator has itself emerged into this world through a process that may be called entropic randomization within the thermodynamic flow, if one wishes to avoid theological language. But the Christian will just as surely want to say of the same emergence of the cocreator that it was through the eschatologically modulated creative action of God.

Furthermore, the quintessential action of the created cocreator, the cultural stretching of the physico–biological reality, is a continuation at different levels of what happens in sex. Or, we might say, of what happens continually within the thermodynamic flow: the attempt to bring the new out of the old, by recombining those elements which have not yet had intercourse, in combinations not yet actualized. Since, as a part of the thermodynamic flow, sex is a dynamic process, the human cocreator is surely changing the multilayered legacy that has been bequeathed to it through evolution.

I entitled this piece "Sex, for God's Sake." What that means may now be clear. Sex, for God's sake, is sex under the purposes of God. What are God's purposes for sex? We have accepted the scientific insight that sex is first of all to be understood as a "means for mobilizing genetic variability."[21] From this, we have gone on to say that perhaps sex is part of God's ongoing work of bringing the new into existence while being faithful to the concrete histories of the creation. God's eschatological structuring of the world is nowhere to be seen more clearly than in the phenomenon of sex. Since this action is not peripheral to God and the divine activity, we might further conclude that what is happening in sex, as discerned by the scientists, is very much part of the mainstream of what the world is all about and what God wishes for the creation.

For all of the above reasons, we may speak of sex as a paradigmatic entity, because it shows forth patterns of meaning that are true of the evolutionary process in its entirety.

PLEASURE, INTIMACY, AND LOVE

Another set of issues must at least be touched upon before we leave the topic. These issues have to do with what the overwhelming number of persons in the general population would have thought about when they heard the word "sex." These are the issues of pleasure, intimacy, love, and, yes, even giving birth—the issues that some Christians speak of as procreation and enhancement of love, and which they consider the chief elements of sex. This conference is clearly not approaching sex from the perspectives that highlight these factors, but we do well, nevertheless, to consider them briefly.

The pleasures of human sex, including female orgasm and male climax, what shall we say of them? I hope that it will not fly in the face of any reasonable understanding to suggest that these are the kind of high-powered motivators that are needed if nearly all the members of the species are to be induced to enter voluntarily into the enterprise of bringing the new into existence by genetic recombination. Scientists have discussed the high cost of sexual recombination. Some even assert that it is only the larger evolutionary units, the species or the clade, that benefit from sex. This being the case, what will induce the members of a thoughtful, self-aware species like the human species to engage in sex? High-powered motivators of pleasure might certainly be part of the answer. But the pleasure motivators we carry within us are not geared to the complexities of human life today. Culture must stretch them also, if they are to be useful. Culture presently is contributing negatively to our society today by stretching these motivators consistently in unwholesome directions.

The pleasures of human sex are not just confined to the act that results in procreation. Perhaps we would do well to introduce the term "sexuality" at this point. The pleasures of human sexuality may not be associated with procreation at all. And even a relationship that focuses primarily on procreation finds that intimacy and pleasure accompany the process. We must also introduce the word that has been almost unspoken here: love. Love and sex have many associations at the human level, and we cannot survey them all here. However, bringing the new into existence through the transformation of the old certainly involves more than genetic recombination. The newborn product of that recombination needs nurture in love. The newborn needs this love for its own survival, but also as the major ingredient of its learning how to be a new creature. It takes nearly thirty years after birth in many societies to produce a body and a trained brain that can function as an adult human being. One of the major things that must be learned in that period of time is that love is essential to the human enterprise.

The transformation of stretching-unto-fulfillment that human culture attempts on the surrounding ecosystem and its inhabitants requires love at its center. Christian faith has believed this from the beginning, and now many others agree. Only love can discern properly how the transformation and stretching should be undertaken. Further, the stretching must lead to an increase of love, else it is not fulfillment.

Love means several things; some cultures, like the ancient Greek, had several different terms to describe the different aspects of what we lump together in English under the term "love." Love is respect for self and others, love is passion and warmth, love is, at its highest, self-giving for others. "Love one another as I have loved you." "Greater love hath no one than this, to lay down one's life for the neighbor." Some scientific observers term this altruism beyond the kin group.[22] They go on to insist that highly complex societies, like human society on the planet just now, cannot endure without the universal practice of such altruism.

Love is not confined to sex and sexuality, but it is very appropriate that it should be associated with these, since the entropic randomizations that we have been talking about cannot possibly succeed in birthing the new in a wholesome manner without love—love both as instrument for bringing the new into existence, and love as the substance of what the new is really about. When we look at the matter in this light, then the physiology and psychology of sexuality, the motivators and the pleasure they give, as well as the enhanced sense of people that is associated with sexuality as its most wholesome—then we can say that at the microlevel within the thermodynamic flow, these motiators and pleasures may contribute to the engines of truest personhood and altruism. This argument could sustain a book in itself.

Nothing has been so subjected to human cultural manipulation and distortion as the dimensions of love that are associated with sex. Such cultural distortions are serious, not least because they represent a perversion of the transforming and stretching function of human culture. They stand as demonic perversions of the essentially good.

My discussion has gone far afield, touched on many areas of knowledge and sensibility. This wide–ranging approach is really quite unavoidable, and it should be carried further. Sex, as I have learned, has first of all to do with the new coming into existence and becoming actual. If this is so, then it stands at the heart of what the evolutionary enterprise is all about, at least in Christian theological terms. Little wonder that when we discuss it, we find ourselves drawn into a consideration of the fundamental nature and destiny of the creation. It could not be otherwise.

NOTES

1. Tradition offers a framework for such thinking, which includes affirmation about creation of this world, human sin, redemption, and fulfillment. Tradition offers us parameters, guideposts for our thinking, albeit they are parameters that are subject to rearrangement and reformulation.
2. Ghislein, M. T. 1974. *The Economy of Nature and the Evolution of Sex.* University of California Press: Berkeley, CA. 53.

3. Margulis, L. 1986. *Origins of Sex.* Yale University Press: New Haven, CT. 9.
4. Goodenough, U. 1985. An essay on the origins and evolution of eukaryotic sex. In: *The Origin and Evolution of Sex.* Alan Liss: New York. 124.
5. Wicken, J. S. 1987. *Evolution, Thermodynamics, and Information: Extending the Darwinian Program.* Oxford University Press: New York.
6. Ibid., 137.
7. Ibid., 84, 218.
8. Ibid., 218.
9. Ibid., 217.
10. Bronowski, J. 1970. New concepts in the evolution of complexity: stratified stability and unbounded plans. *Zygon: Journal of Religion and Science 5* (March):18–35; Denbigh, K. 1975. *An Inventive Universe.* Braziller: New York; Katchalsky, A. 1971. Thermodynamics of flow and biological organization. *Zygon: Journal of Religion and Science 6* (June) 99:125.
11. Wicken, *Evolution, Thermodynamics, and Information,* 218.
12. Ibid., 8.
13. Ibid., 214.
14. Ibid., 213.
15. Ibid., 218.
16. Ibid., 219.
17. Burhoe, R. W. 1976. The source of civilization in the natural selection of coadapted information in genes and culture. *Zygon: Journal of Religion and Science 11* (September):263–303; 1979. Religion's role in human evolution: the missing link between ape–man's selfish genes and civilized altruism. *Zygon: Journal of Religion and Science 14* (June):135–162. 1986. War, peace, and religion's biocultural evolution. *Zygon: Journal of Religion and Science 21* (December):439–72.
18. Csikszentmihalyi, M., and F. Massimini. 1985. On the psychological selection of bio–cultural information. In: *L'Esperienza Quotidiana: Teoria e Metodo d'Analisi.* F. Massimini, and P. Inghilleri, eds. F. Angeli: Milan, Italy; Csikszentmihalyi, M. 1986. Cultural evolution and human welfare. Unpublished paper, delivered at the annual meeting of the American Association for the Advancement of Science, Chicago, Illinois, February 1987.
19. Hefner, P. 1987. Freedom in evolutionary perspective. In: *Free Will and Determinism.* V. Mortensen and R. Sorenson, eds. Aarhus University Press: Aarhus, Denmark.
20. Theissen, G. 1985. *Biblical Faith: An Evolutionary Approach.* Fortress Press: Philadelphia, PA.
21. Ghiselin, M. T. 1974. *The Economy of Nature and the Evolution of Sex.* University of California Press: Berkeley, CA. 57.
22. Campbell, D. T. 1975. On the conflicts between biological and social evolution and between psychology and moral tradition. *American Psychologist* 30:1103–1126; Burhow, R. W. 1979. Religion's role in human evolution: the missing link between ape–man's selfish genes and civilized altruism. *Zygon: Journal of Religion and Science 14* (June):135–162. 1986. War, peace, and religion's biocultural evolution. *Zygon: Journal of Religion and Science 21* (December):439–472.

PANEL DISCUSSION

AUDIENCE QUESTION: I would be very interested in hearing how Dr. Maynard Smith would use the idea of entropic randomization to explain the present–day success of sex.

MAYNARD SMITH: A short answer is that I do not know how it does. More concretely than that, I do not think that any approach using the second law of thermodynamics has anything at all to do with science. That may seem very strange since thermodynamics, is, as a matter of fact, an important branch of physics. But you see, to be something to do with science, a theory has to do some work. It has to lead to some consequences of some kind, some predictions. I really do not think that any of the applications of thermodynamics to evolution have that property. I think that there is something sort of deeply worrying here. We all probably differ to some extent in our explanations of the function of sex, but I think we are all searching for the same kind of explanation of sex. We all want to explain what we see causally, and that means explaining the present in terms of the past. If I understood Dr. Hefner correctly, his explanation of sex was, to use his word, "eschatological." It was to explain sex in terms of God's purpose for the future . . . to explain the present in terms of the future. I don't think these two approaches can be made compatible.

HEFNER: Do you think the one can be consistent with the other? Could the theologian make statements that are more or less consistent with what the scientist understands?

MAYNARD SMITH: I do not want to be understood to say that I don't think there is a role for anything other than science. Let me be quite explicit, I do not think that scientific statements are the only kinds of statements that people have to make. If you want to tell a woman that you love her, it is not a scientific statement. What worries me is that we will mix these kinds of statements together and produce confusion. I felt that Dr. Hefner was confusing these two kinds of language.

MARGULIS: In spite of the potential for confusion, I would like to laud Dr. Hefner for emphasizing the role science can play in developing theology. Certainly there are many people who feel exactly the opposite: that science simply interferes with theology, and in fact leads to nothing but nuclear weapons. I can understand that view, but I hear Dr. Hefner saying that science is a way of knowing, and it's even a way of shedding light on issues that have been in the tradition of theology.

HEFNER: The two ways of knowing, science and theology, must be brought into consistency. They are not going to be brought into consistency by tinkering with science. They're going to be brought into consistency by doing something with theology.

MARGULIS: I would like to bring up another issue. I couldn't help but get the impression that you were saying that as a "created co-creator" humans are above apes, better than apes, and that the earth is here for our use . . . to quote Dr. Hrdy, quoting Katherine Hepburn, that "nature is what we are put in this world to rise above." I see this as a dangerous and erroneous view.

HEFNER: I certainly agree with that. I hope that I wasn't saying that in any simplistic way. It's difficult linguistically to differentiate between what you have just brought up and the notion that nevertheless, human culture can do and does do things with, to, and for nature in a way that no other creature does. There comes a responsibility to follow some sort of criteria of wholesomeness. Of course in saying that, I am using a version of Katherine Hepburn's famous line.

MARGULIS: Well, I see man—people rather—as great mammalian weeds in the sense our human populations grow quickly and alter the environment beyond recognition by anything but cockroaches, rats, coke bottles, and beer cans. If you want to laud human culture for these "achievements," we are in big trouble. On the other hand, if your message is one of concern that values and judgments are intrinsic parts of knowledge, then I am very sympathetic. Scientists are deficient in recognizing knowledge to be value–laden. Our communication problems may be ones of vocabulary, as Dr. Maynard Smith suggested. The scientist's vocabulary shouldn't even include "creature," for example, as the creatures you speak about are the products of creation. The last thing in the world I believe is that we are the products of direct creation.

HEFNER: I can understand that, and what Dr. Maynard Smith said as well. Perhaps there is an unavoidable appearance of putting things into a comprehensive whole, and of course you can't speak of a comprehensive whole with scientific finality. That is when we get into the realm of myth, poetry, and theology. If the will of God is interpreted in a perverse way, as it has sometimes been in Christian history, where the purpose of nature was as something for humans to twist to their advantage, then of course you are in demonic trouble. If, however, the understanding of God and the will of God is interpreted in a beneficial and life–giving way, then it becomes one of the most powerful motivators that we have.

HRDY: Since this is a theological discussion, I want my turn to explain the terminology. I think Dr. Margulis is exactly right in her interpretation of Katherine Hepburn's phrase "rise above." Even so, I want to emphasize an even more subtle point. I think it is safe to say that at least four of the five people on this panel (maybe more, but I do not want to implicate the innocent unnecessarily), don't believe in God and probably don't believe that humans are inherently more valuable than other species. However, I think it is also safe to say, though I am overassuming their innocence, that these same panelists essentially act in line with Christian standards of behavior and certainly in line with civilized standards of behavior. I consider myself a Christian in action, even without my believing in God. We are not behaving like monkeys, and what a miserable existence for anyone who succumbs to that fallacy. If there is anything we can do to make our lives less than "nasty, brutish, and short," it is to abide by these standards of behavior. It is those standards that I want to hear more about from theologians. What is a good and decent way for humans to live, given the truly unfortunate heritage that we come onto this earth with?

HEFNER: A theological presentation like mine is aimed more at understanding than at ethics. It isn't going to satisfy that particular desire of yours. that is a realm of theology all its own.

AUDIENCE QUESTION: The central idea of Christian theology, the incarnation, seems to deny your argument that sex is God's way of doing business in the world.

HEFNER: If you took my rather abstracted notion that sex is a stormy liaison that results in recombination that may be beneficial, then I think, in the abstract, the principles of Christ fits in that same paradigm. Christ, in the life of Jesus of Nazareth, is entropic randomization. There was a clash of cultures, a clash of religious perspectives, a clash of ideas over how society should be maintained, of how slave and free, Jew and Greek, men and women, should interrelate. If that was not entropic randomization, I do not know what is. There was a very stormy process, and people who don't accept the Christian claims about Jesus think that it was a process of death and defeat. The Christian perspective is that it was the birth of a new, a resurrection.

AUDIENCE QUESTION: Does evolution by natural selection drive a biological system to a morally acceptable state?

HEFNER: You can't really speak about a biological system being driven to a moral state. You have to bring in culture. Sometimes it is said that the cultural system should conform to the biological. In other cases, we say that the cultural system should do all it can to negate the biological system. It is a question of values, a topic for another conference.

QUESTIONS FOR FURTHER DISCUSSION

1. Does Hefner say that the randomness of sex is the realm of God, and that the part of sex that is explained by evolutionary theory is the realm of science?
2. Hefner deliberately did not address ethics or moral guidelines in his address. Why not? Are there no moral guidelines to be drawn from biology?
3. What separates the approach of the scientists from that of the theologian? Can the two views be reconciled?
4. Why shouldn't we tinker with science to "bring it into consistency" with theology, yet tinker with theology to make it consistent with scientific fact?

If you have followed the progress of the book to this point, you may consider this final chapter to be out of context. Not so. We have taken you from the evolution, persistence, and diversity of sex to a theological perspective. What remains is to show one final twist of how sex—sex differences actually—can have a direct influence in an area where you might least expect it, in a discipline that is popularly presumed to be objective, free of prejudice and bias: science itself. Dr. Hrdy in the concluding address delivered after the Nobel Conference dinner, shared her perception of how research has had a male bias. The panel discussion at the conclusion of her talk was lively and provocative. We share some of it with you to give you the flavor of the evening.

The Editors

7. Raising Darwin's Consciousness: Females and Evolutionary Theory

SARAH BLAFFER HRDY

When people first began seriously to study the behavior of monkeys in their natural habitats, attention of the researchers gravitated to the behavior of adult males. Among most of the group–dwelling terrestrial monkeys (those easiest to study), there were virtually always fewer adult males than females. These males were much larger than females, and their behavior was more boisterous. Male behaviors were more conspicuous, and males were easier to recognize as individuals. But there was more to this research than just a male–oriented focus, for the observational and methodological biases came linked to biases of much older standing—dating back to Darwin, to the nineteenth century generally, and to even older antecedents. Among other things, researchers were enthralled with a powerful theory: Darwin's theory of sexual selection. According to this theory, males actively compete for access to females. In the course of this competition, the stronger male prevails, dooming his rival to relatively fewer reproductive opportunities than the winner will enjoy. Competition between males then led to selection of bigger and more muscular males, so that in the famous example of the Hamadryas baboons, males evolved to be nearly twice as large as females belonging to the same species. Male hamadryas baboons are not only bigger, but far flashier in appearance, endowed with an intimidating mane of hair and a face the color of raw beef steak—as different from the mousey grey–brown females as if they belonged to two different species (see fig. 1). Male–male competition was half of Darwin's theory of sexual selection; the other half had to do with female

Figure 1. Hamadryas baboon males from two different groups fight while the females huddle in the background. These baboons are one of the very few primate species that actually do live up to the old stereotypes of agressive males and passive females! (Photo courtesy of Joseph Popp, Anthro–Photo.)

Figure 2. Hamadryas baboon male herds a female. The males are roughly twice the size of the females. (Photo courtesy of Joseph Popp, Anthro–Photo.)

choice, the notion that females by nature will seek to select the single best male as a breeding partner from out of a panoply of competing suitors. As a matter of fact, this part of the theory does not apply very well to monkeys, and particularly not the hamadryas baboon due to certain peculiarities of its breeding system. A female hamadryas baboon is adopted while still a juvenile by an adult male on the make. The male will herd her about for the rest of her life, nipping her on the neck to assure her proximity (see fig. 2). But forget those details for a moment, and focus on the central assumption, that males play the more active role. For the male hamadryas baboon, with his lion's mane and his muscular and domineering disposition, provides the perfect model of a modern sexually selected male. It also happens, however, that the hamadryas case is virtually unique among primates, the only case out of some 175 extant species of primates where we can actually find any sort of clear–cut dichotomy between competitive males and passive females! Instead of the patriarchal hamadryas case, we could just as easily have focused on any of a number of lemur species, species in which females rather routinely dominate males. We could have decided to make an example of the shy and nocturnal owl monkey (*Aotus trivirgatus*), where males and females cooperate in child care with the male playing the major role in carrying and protecting the infant, or we could have focused on the gentle South American monkeys known as "muriqui" (*Brachyteles arachnoides*), who specialize in *avoiding* aggressive interactions, or any of a host of other primate species in which we now know that females play an active role in social organization. But the history of primatology did not unfold that way. Instead, until very recently, a hamadryas–like stereotype was taken as the primate norm.

In retrospect it is remarkable that we ever could have believed that selection primarily operated on only one sex, and yet that is precisely the assumption that until recently did underlie many conclusions about primate breeding systems. Consider the treatment of nonhuman primates in a recent textbook in sociobiology. The author describes how male monkeys, such as rhesus macaques, compete among themselves for access to females so that only 20 percent of males are responsible for 80 percent of the breeding, while *all* the females that come into estrus tend to be impregnated. "These data make it clear that only males are directly involved in differential selection among rhesus [monkeys] and probably all the terrestrial and semiterrestiral primates."[1]

A cluster of biases, then—methodological, ideological, and theoretical—contributed to an extraordinary phenomenon: an intellection formulation of primate social organization that lasted for over twenty years and that—based on what we know today—was totally unsupportable. This of course is not the first time that social preconceptions have caused scientists to seriously misinterpret nature, but it is one of the more clear–cut and better documented examples.

To continue this story, but still keep it simple, I will stick to baboons. Let us shift then from the patriarchal hamadryas to a closely related cluster of species known as savanna baboons, which instead of living in harems lives on African savannas in large, multimale troops. These savanna baboons were the first monkeys to be extensively studied, and they were depicted as having a social structure that in many respects was the mirror image of the kind of organization then found in American universities and corporate structures. There was a central male hierarchy in which competing adult males formed alliances with other males in order to maintain high status. Female baboons were viewed as pawns in this game, and sexual access to females was the reward for males successful in maintaining high rank. Whereas males were thought to have almost nothing to do with infants, females were thought to be so absorbed in child care that they had almost no impact on the social structure of the group. What was missing, of course, was any emperical description of the full range of activities of either sex. Lionel Tiger summed up the prevailing opinion: "Primate females seem to be biologically unprogrammed to dominate political systems, and the whole weight of the relevant primates' breeding history militates against female participation in what we call 'primate public life.' " Yet once we begin to examine the actual evidence, few statements could have been further from the truth for any species, except just possibly the hamadryas baboon.

Let's take a closer look, then, at the species which has become anthropology's "type case" for a male–dominated social order. What happened when we identified females as individuals and monitored their behavior over time? The picture changed radically.

The main difference between savanna baboon males and females is not that males are active and females passive, but the fact that females stay in the troop of their birth while males are transients. A male moves every four or five years, and within the troop, his status is in perpetual flux. Typically, a young male leaves his natal troop about the time he matures, and attempts to enter another through a gradual process of insinuating himself into the group. Sometimes a males does this by first forming a friendship with a troop female, who serves as a sponsor for his membership in the group. Male–female friendships are not so much based on dominance as on mutual interactions, such as grooming, in which either sex may take the initiative. That is, not

Figure 3. Recent studies of savanna baboons reveal highly developed longterm relationships between females and males as between these two *Papio anubis* "friends." (Courtesy of Barbara Smuts, Anthro–Photo.)

all males are fighting their way into the troop by allying with and defeating other males. Indeed Barbara Smuts who describes male–female friendships in detail in her recent book *Sex and Friendship Among Baboons*, tells a wonderful anecdote about a female who enters a neighboring troop to lure back with her a particular male to which she seems to have taken a fancy, initiating his entry into her own troop (see fig. 3).

Earlier studies that focused on male–male competition for breeding access to females gave us a very skewed picture indeed. Invariably, researchers focused their attention on things like counting the number of copulations for the males so that male-–female interactions were usually only recorded when the female was in heat. Instead, Smuts focused on females in all stages of their reproductive cycle. Her analyses revealed that females select and preferentially stay near one or two of the eighteen or so adult males in the troop, and these relationships remain constant through pregnancy and lactation.

Not only are male–female relationships much more reciprocal and complex than previously realized, but there is also much more involvement by males with infants. Once a female baboon gives birth, one or several of her male friends provide various babysitting services for the mother. In terms of actual time spent with the infants, it's rather like the human case: not much. That is, if you are standing on the savanna watching a troop of baboons, you'll see about one male–infant interaction once every ninteeen hours. However, the protection offered just by the proximity of these males may be critical for infant survival—particularly for discouraging attacks on the infant either by incoming males who are unfamiliar with the infant's mother, or harassment of the infant and mother by females from competing lineages in the troop.

Once we understand the importance of male involvement with infants, the internal politics of a baboon troop take on new dimensions. Female baboons, for example, actively engage in forging for themselves a network of alliance with different males. In short, there is much more going on than simply males competing with other males. Males are maneuvering for access to females, while females themselves are busily building alliances with males. Both sexes of course are also preoccupied with survival, keeping safe, staying fed, and this leads to another very important set of female activities. Females cooperate with their relatives, their mothers, and grandmothers, in order to compete with females in other matrilines belonging to their same troop. Competition is for such things as resources and what might be called "living space" or freedom from harassment. The resulting structure from these various female preoccupations turns out to be remarkably persistent and stable.

When two females of different social status approach each other (a dominant female approaches a subordinate or vice versa), you are likely to witness a remarkable performance. The subordinate female greets a dominant female by presenting to her; giving an exaggerated "fear grin," lifting her tail, and jerking a foot back. Even more remarkable is the fact that in an episode like this we can be fairly certain that the main reason this female is dominant is because her mother was.

Not only does there exist a stable hierarchy among females, but it is a very conservative hierarchy, predictable from one year to the next and even one generation to the next; so if baboon social structure is to be understood, relations must be understood.

To make a long story short, then, we completely failed to recognize first the many very active roles females were assuming in troop affairs, and second the many other things—like caring for offspring–that males do.

The collection of data on female behavior from a wide range of species (such as tamarins, lemurs, and woolly spider monkeys) has not only caused us to revise our notion of female nature—to encompass creatures that *are* nurturing—but that are also aggressive, competitive, cooperative, and a wide range of other things. But such data have also forced us to reinterpret the behavior of males.

We have been forced to expand our theoretical constructs to incorporate the full range of selective pressures on both sexes. The assumptions underlying such revised theory are very different from the earlier formulations. For example, by shifting our focus from the production of infants to the survival of infants, we are forced to take account of a whole range of male and female activities that have drastic repercussions on the survival of offspring.

So much for raising of Darwinian consciences. What about the scientific endeavor generally? I have documented just one example of how, for over two decades, researchers in my own field completely misconstrued primate breeding systems because of such bias. The real question is, just how damaging is this?

It seems to me that documenting these biases and starting to look at the world from a female point of view has been terribly valuable in revising history, literature, and even primatology; but it has also contributed to a growing cynicism about science generally, and especially social science. By pointing out the pervasiveness of preconceptions and biases in virtually all scientific and scholarly endeavors, feminist scholars have contributed to a general and quite fashionable challenge currently hurled at science. Given that all scientists are embedded in their cultures and that all research is inevitably informed by cultural bias, the question they ask is: "Can we really *know* anything?"

Clearly, it is unacceptable to permit old biases, once discovered, to persist. It is undeniable that most fields, including history, psychology, and biology, have been male–centered; but the noteworthy and encouraging thing is how little resistance there has been to revisionist enterprises once begun. On the contrary, in fields like sociobiology, there has been something more like a small stampede to study female reproductive strategies so that there exists a real danger that we will now merely substitute a new set of biases for the old ones. According to one emerging revisionist dogma, for example, it is now finally acceptable to say that men and women are different, provided we also specify that women are "cooperative, nurturing, and supportive," not to mention equipped with unique moral sensibilities. Entering the fray from a different perspective, various religious sects would also like to benefit from the current disarray to inject their agendas. Yet there can be no advantage for any scholarly enterprise to specify what can or cannot be found.

In spite of its limitations, scientific inquiry as currently practiced, with all of the drawbacks—including reductionist models, underlying assumptions that have been influenced by cultural context, domination of disciplines by males, and so forth, all the things that gave us several generations of male–biased primatology—science with all these drawbacks is still better than such unabashedly ideological programs that have become advocated in certain religious as well as in some feminist research programs (such as those advocating "conscious partiality"—the notion that since we can't help being biased, let's be biased in an ideologically correct way).

Needless to say, I reject such programs. I accept that the best we can do is to try to remain intellectually independent, to invite multiple inquiries, and to encourage restudies and challenges to current theories. Essentially, then, this is science as currently practiced—inefficient, replete with false starts in need of constant revision—but still better than any of the alternative programs being advocated.

NOTES

1. Freedman, D. 1979. *Human Sociobiology.* Free Press: New York. 33.

PANEL DISCUSSION

MAYNARD SMITH: There is a temptation for males to make the comment that, "Well, Darwin was a Victorian and with that background was bound to be somehwat of a male chauvinist, but I'm alright, you know, I'm a decent chap." The following rather sad anecdote about my own scientific career occurred to me. It's trivial, but it's nevertheless quite revealing.

When I was just a beginning scientist, I worked with the fruit fly *Drosophila* which I loved greatly. I was interested in the causes of aging. Why do animals die of old age? One of the things I discovered was that the actual process of laying eggs causes female fruit flies to die sooner than anything else you do to them, even giving them large doses of radiation that stop them from laying eggs. I published this observation and got a lot of credit for it. At the time it never occurred to me to ask if mating could have an effect, good or bad, on the longevity of males. Twenty years later, a young scientist demonstrated that if you allow a male to mate twice a day, he dies young. I don't mean to give you more than one guess as to the sex of the scientist that made this discovery. It was my friend Linda Partridge.

I could have done that experiment painlessly. I didn't say to myself that it would be terrible to discover such a thing and that "I won't do it." So those of us who are absolutely sure that we have no prejudices are almost certainly wrong. It may well be that the prejudices we have, we have because we're men and most scientists are men. They are not, of course, the only prejudices we have, because our color or race can be just as important in how we see the world.

However, the thing that I agree most passionately with was what Dr. Hrdy said towards the end of her lecture. It was this: "Okay, so we are prejudiced. We can try not to be, but we are not going to succeed. We can never be free of our preconceptions." I think there is an obvious way in which we can try to avoid our prejudices and that is to recognize that science is not an activity carried out by a single individual human being. It is a collective activity.

It has been nice to be at this conference with two women who can tell the rest of us to shut up occasionally, and not that it has any effect on me, but it has an effect on others. I think that if we can ensure a sufficiently mixed group of people doing science, we can't get rid of all our prejudices, but at least we will reduce them.

HEFNER: Dr. Hrdy deals primarily with biases in how data are collected and analyzed. I have heard a lot about the differences between male and female thinking in science, and I wonder if you feel this has an eroding effect on the credibility of science. You know, "Science is just another form of mythology, not much different from theology." "Why should we listen to the scientists?" And so forth.

HRDY: I think that there has been an erosion in confidence in science. This notion that if all knowledge is relative and everybody is biased, then we can't know anything. The very extreme fringe of the creationist movement takes advantage of those fears to say, "Look, they don't know things any better than we do." I think we are recognizing our fallibility, something the church has been telling us for years.

This objectivity is going to prove to be elusive, and the best we can do is to include in science not just people of both sexes, but also minorities, people who have been oppressed, who know what it feels like to be oppressed, and so forth. Let's say you're studying primate behavior and you are interested in dominance hierarchies. Just how critical is to have someone out there who can identify with the organisms they are studying? For example, the Japanese spend inordinate numbers of pages writing about the phenomena of social ostracism among monkeys. To my knowledge, few American or British primatologists have ever even concerned themselves with the topic of ostracism.

QUESTIONS FOR FURTHER DISCUSSION

1. The scientists have a vested interest in trying to appear unbiased. Have they overstated the self–corecting mechanisms "inherent" in the scientific process?
2. Are there some types of research questions that should not be asked because of the impact their apparent solutions would have on human society?
3. Can you imagine situations where a male researcher would ask different kinds of questions than a female investigator, when the research topic had nothing to do with sex?

Glossary

Entries into this glossary were made with reference to (1) Nowak, R. M., and J. L. Paradiso. 1983. *Walker's Mammals of the World*. Johns Hopkins University Press: Baltimore, MD; (2) Wallace, R. A., J. L. King, and G. P. Sanders. 1981. *Biology, The Science of Life*. Scott, Foresman and Company: Glenview, IL; and (3) Scagel, R. F., R. J. Bandoni, J. R. Maze, G. E. Rouse, W. B. Schofield, and J. R. Stein. *Plants, an Evolutionary Survey*. Wadsworth Publishing: Belmont, CA; and represent current usage of these terms. Further information can be found by consulting these or other references works.

Abiotic   With life or dead.

Adaptation   An evolutionary modification that better suits an organism to its biotic and abiotic environment.

Adaptationist   One who argues that many (if not all) of the observable traits of organisms are adaptations, (that is, they have been shaped by natural selection for their present roles).

African vervet monkeys   Members of the species *Cercopithecus aethiops*, found in Sudan and Ethiopia, whose close relatives include the green monkeys and guenons. They are predominately fruit–eating omnivores of less than 9 kg. in weight, whose social structure is poorly known.

Allele   Alternative character states for a given gene locus. For example, the eye–color gene has several alleles: blue, green, brown, and so forth.

Allen's swamp monkeys   Called *Allenopithecus nigroviridis* by primatologists. This species is found only in northeast Congo and northwest Zaire. These monkeys forage on the ground much like baboons, making them relatively easy to study, but little is known of their social habits.

Alouatta   This genus is composed of the howler monkeys, arboreal leaf–eaters of Central and South America. Their habit of roaring at dawn has earned them their name. In general, they move in large, multimale groups of up to forty–five individuals, usually dominated by a single male.

Alternation of generations The life cycle of many plants and animals, which switches from haploid to diploid and back to haploid, and so forth, with the transition to the next generation.

Altruism Selfless behavior; its more restrictive definition when used in a biological context is behavior that is beneficial to non-related individuals, while detrimental to self.

Amborellacea An obscure family of primitive angiosperms.

Androgynous Being both male and female at the same time; or, in the case of plants, having flowers with both male and female parts.

Angiosperm Any member of a large taxonomic grouping of plants that have their eggs encased in ovaries. This group contains all plants one associates with showy flowers.

Anogenital Pertaining to the region of the body between and around the anus and the genitalia (synonymous with perineal).

Anther The pollen–producing structure of flowers.

Antheridium The male structure of primitive plants that produces sperm (plural: antheridia).

Anthropologist One who studies people.

Antibiotic Any chemical substance that kills or inhibits the growth of an organism.

Antigens The substances that stimulate the immune system to mount an immune response to a foreign invader.

Anus The terminus of the digestive tract through which fecal material passes.

Aphrodiasiac A substance that enhances the pleasure of or increases the likelihood of sexual activity.

Arboreal Spending much of one's life in trees.

Archegonia A multicellular structure of primitive plants that produces the female gamete, the egg (singular: archegonium).

Arthropods A large taxonomic group of organisms, including the insects, crabs, and so forth.

Asexual Reproducing without sex.

Aspergillus A mold similar to bread mold.

Australopithecus afarensis A fossil hominid first discovered by Raymond Dart in 1924. They lived more than 3 million years ago and were of short stature (about four feet tall), had human-

–like teeth, and walked on two legs. There is some evidence to suggest they were hunters and used bone tools to kill.

Autocorrelation A measure of the tendency for past events to predict future ones. Used to describe the period length of cyclic histories.

Automixis The fusion of an egg with one of the polar bodies to reestablish the diploid state and then allow development of the egg into an adult.

Barbary macaque Any member of the species *Macaca sylvanus*, from northwest Africa. Macaques lack the long tails of many of the other monkeys; the Barbary macaque is extreme in being completely tailless. These macaques are very generalized in their habitat requirements, and use a variety of areas—lowlands, upland areas, even altitudes over two thousand meters. They are largely vegetarian. Multimale groups can become very large (greater than 100 individuals).

Bennittitales An order of plants (also called the Cycadeoidales) that reached dominance during the Mesozoic period.

Bilaterally symmetrical Having an irregular organization such that only one vertical plane divides the structure into two mirror images (for example, humans are bilaterally symmetrical with a plane drawn from nose to navel). See *radially symmetrical.*

Biomass Pertaining to the weight of tissue derived from living or once–living organisms

Bipedal Having two legs.

Black colobus monkey Also called *Colobus satanus.* Found in several areas of equatorial West Africa. These monkeys are deep forest residents, eating leaves and flowers. Little is known of their social structure.

Blue monkey Known as *Cercopithecus mitis* by primatologists, these monkeys weight up to twelve kg. and live in the forests of central East Africa. In general, the monkeys of this genus have elaborate beards and cheek pouches. Their diet is primarily fruit and other plant parts, but they also eat baby birds, lizards, insects, and so forth. Their troops contain between thirteen and twenty–seven individuals, several of which are adult males.

Botany The study of plants.

Bottleneck A situation where a population, through some form of external disturbance or through migration, is reduced to a very small number of individuals. This condition, if frequent, tends to reduce the heterozygosity of the population even after it recovers from the disturbance.

Bryophyte   A moss, liverwort, or hornwort.

Bulbil   A small bulb produced through the mitotic cell divisions (for example, as seen in onions or lilies).

Carotenoid   Any of a class of orange–red pigments used by plants in the photosynthetic process. These pigments are what give trees their beautiful fall colors.

Carpel   A specialized leaf modification that surrounds the ovule and forms one of the petals of some flowers.

Caytonia   An extinct seed fern of the Mesozoic.

Celebes macaque   Known as *Macaca maurus,* by primatologists these stocky–bodied, stubby–tailed relatives of baboons are found only in southwest Celebes, Indonesia. Groups vary in size from five to twenty–five individuals. Macaques, in general, are primarily fruit–eating omnivores.

Cell plate   A rigid structure that divides daughter cells formed during cell division in plants.

Cellulose   A sugar produced by plants for structural support; it is indigestible by most animals.

Centriole   The microtubule anchor that supports the apparatus that allows for the mechanical separation of chromosomes during cell division.

Cercopithecine monkeys   This family of monkeys is composed of eighty–five species found principally in Africa, southeast Asia, and Japan. Their tails are not prehensile. There social structures are quite variable.

Cercopithecus   The genus of monkeys that includes the blue and vervet monkeys. All members of this genus are long–legged, long–tailed primates called Guenons. There are twenty species in the genus, all from equatorial Africa.

Charophyceae   A family of green algae containing approximately 250 species that remain attached to the bottom of freshwater and brackish bodies of water.

Chiasma   The site of a crossing over event in meiosis.

Chlamydomonads   Referring to the members of the genus *Chlamydomonas*

Chlamydomonas   The genus name of a single–celled self-propelled water plant with a single light–sensitive eyespot. They are found in freshwater ponds everywhere.

Chloranthaceae   A family of green algae.

Chlorophyll   The light–energy abosrbing pigment that is used to capture photons during photosynthesis.

Chlorophyta   Referring to a mixed collection of green algae species thought to have given rise to the land plants.

Chloroplast   A membrane–bound structure found inside plant cells, whose biochemical function is to carry out photosynthesis.

Clade   A lineage, usually meaning a group of related species or genera.

Cladoceran   Any of a taxonomic grouping of microcrustacea, typified by the water fleas.

Clone   A lineage of cells or organisms produced without the aid of meiosis. There is no mixing of genetic material from outside sources in the offspring of clones.

Clubmosses   Any plant of the genus *Lycopodium*, the ground pine.

Cnemidophorus   The genus that contains the whiptailed lizards.

Cnemidophorus uniparens   Also called the desert–grassland whiptail; its native range is the southwestern United States. A handsome brown– and yellow–striped, long–tailed insectivorous and parthenogenetic lizard.

Coadapting   Interacting organisms evolving with regard to their shared interactions.

Coevolution   The pairwise evolution of the two interacting organisms, such that one evolves in response to changes in the other, such that the other then evolves in response to changes in the first, and so forth.

Coevolve   To evolve in a coevolutionary fashion.

Coleochaete   The genus name for a group of photosynthetically active organisms that have a filamentous growth form composed of chains of single cells. All members of this genus are aquatic. It is currently thought to be part of the algae lineage that did not lead to the land plants.

Colobine monkeys   Referring to that subfamily of primates containing the langurs, colobus monkeys, and proboscis monkeys. Members of this subfamily are found primarily in Africa,

India, and southeast Asia.

Colobus A genus of equatorial African leaf–eating monkeys containing ten species.

Common chimpanzee Primatologists call this species *Pan troglodytes.* The chimp that finds its way onto our television and movie screens originally comes from central Africa, where it spends most of its days searching for fruit, tender buds, seeds, insects, and meat. The sizes of chimpanzee groups can reach eighty individuals, but they usually fragment into numerous subgroups composed of several adult males and females.

Complementary resistance The condition where resistance to one pathogen leads to susceptibility to another, and vise versa.

Conjugation Joining for the purpose of exchanging genetic material, said of bacteria.

Copulable Having the ability to copulate.

Copulate To engage in sexual intromission; this coupling need not lead to pregnancy.

Corystosperms An extinct group of gymnosperms present in the Mesozoic.

Crab–eating macaque Primatologists know this species as *Macaca fascicularis.* They are found wild in southeast Asia, Indonesia, the Malay Peninsula, and the Phillippines. Their name is derived from their habit of foraging in the intertidal zone. Considerable research effort has been expended in attempts to understand their male–centered social dominance hierarchy. Groups have been reported to be multimale and to contain up to one hundred individuals.

Crossing over The formation of exchanges of gene loci during meiosis. This is the process that produces recombination.

Cuticle The outer, waxy coat of a plant, or the surface of the exterior of insects.

Cycad A palm–like gymnosperm presently restricted to tropical latitudes.

Cytogenetics The study of the behavior of chromosomes within cells as it relates to cell functions.

Darwinian selection Natural selection.

Dehiscence Shedding.

Demography The statistical study of populations.

Desiccate To dry out.

Desmids Simple organisms similar to diatoms, some of which deposit calcium carbonate in their skeltons.

Dialysis machine A device used to filter and thereby clean the blood of people whose kidneys have failed.

Dichogamous Using temporal separation of the maturation of pollen and eggs to avoid self–pollination.

Dioecious The condition of producing only flowers of one sex on a single individual plant. Male and female individuals exist when the species is dioecious.

Diploid Containing two copies of each gene locus carried on paired homologous chromosomes, one from each parent. To be distinguished from the haploid state, where each cell contains only one copy of each gene locus.

Diploidy The state of being diploid.

Directional selection Selection favoring the extremes of a frequency distribution of a given character state (for example, selection for the bluest eyes, the broadest nose, and so forth).

Disjunction Separation of chromosomes during cell division.

Distal More to the extremity, further away from the center. Opposite of proximal.

DNA Literally, deoxyribonucleic acid, the double–stranded genetic material of many (but not all) organisms.

Drill Also called *Mandrillus leucophaeus,* found in Nigeria and Cameroon. The primate is well known as a result of its spectacular facial color pattern. The nose is paralleled by ridges of blue and purple, which make for a nice photograph when viewed face–on. This species is all brown, and is easily distinguished from the mandrill, which has a green coat with yellow undersides. Drills feed on plant material and small animals. They live in groups dominated by a single male, which frequently fuse to form large assemblages.

Drosophila The genus of flies that includes the fruit fly; favored by geneticists for their laboratory investigations.

E. coli, *Escherichia coli,* a species of bacterium that lives in the human digestive tract and elsewhere.

Ecological Pertaining to the interactions an organism has with its biotic and abiotic environments.

Ecologist One who studies the ecological aspects of organisms.

Ejaculate To eject or dispense, usually referring to sperm.

Electrophoresis   A process whereby the chemical composition of individual cells can be used to score genetic markers carried by the individual organism. Slight differences in the migration rates of enzymes in an electrical field are used to search for genetic differences between individuals. This process has revolutionized genetics because it allows the rapid determination of the genotype without elaborate test crosses.

Electrophoretic   Pertaining to electrophoresis.

Embryo   The multicellular developmental stage formed after the fusion of genetic material, during which major morphological and physiological changes occur that lead to adulthood.

Embryogenesis   Formation and development of the embryo through its various stages to the adult.

Embryophytes   Plants that have a multicellular stage of development, the embryo, that stays attached to the parent for at least a part of its development.

Endocrine   Pertaining to the production, action, or distribution of hormones.

Endomitosis   A modification of the process of gamete formation, where an extra duplication of chromosomes occurs before the typical meiotic process. Rather than producing haploid cells, as in meiosis, endomitosis produces diploid cells, since it began with four copies of each gene locus rather than the normal two copies.

Endosperm   The storage tissue found in seeds.

Enteric   Pertaining to the intestines or digestive tract.

Entropic   Pertaining to entropy.

Entropy   The amount of energy in a closed system that is not available to do work; also defined as a measure of the randomness or disorder in a system.

Epithelial   Pertaining to the cells of the skin.

Eschatological   Pertaining to that branch of theology that deals with the final events in human history: death, the resurrection, immortality, and so forth.

Estrus   The period of time when female mammals are fertile, or are in heat; the adjectival form is estrous.

Ethology   The study of animal behavior.

Eukaryote   Organisms whose cells contain membrane–bound internal structures. To be distinguished from prokaryotes (for example, bacteria), all of which lack such structures as mitochondria and endoplasmic reticula).

Evolution   A genetic change in a population through time.

Fertilization   The fusion of cells leading to the mixing of their genetic material.

Fitness   Short for relative fitness, the propensity of an organism for reproduction and therefore representation in the next generation.

Fitness landscape   A mental construct used to illustrate the relative success of different genotypes in a population. Imagine a 3–D plot of fitness on the vertical axis, and different allelic states on the two horizontal axes. Peaks in this 3–D figure are combinations of alleles of high relative fitness. Mine shafts are positioned at the coordinates defining genotypes that show dismal failure in reproduction.

Fixation   The state of allele frequencies where one allele completely dominates the population such that no other alleles for the given gene locus are found.

Flagella   Whip–like structures used to power the movement of sperm (singular: flagellum).

Floaters   Individuals who are excluded from reproduction and from high social standing because they have lost contests with socially dominant individuals.

Formosan macaque   *Macaca cyclopis* is the name given to this species from Taiwan. Presently it is restricted to upland forests by human agriculture, and very little is known of its social structure. Its near relative, the Japanese macaque, is frequently found in zoos.

Fungi   A taxonomic grouping of plant–like organisms that lack chlorophyll (for example, mushrooms, yeast).

Galago   A genus of squirrel–sized primates (less than 2 kg.) from Central and South America. They are known for their large eyes and long bushy tails. Although their fingers do not have suction–cup tips, the broad ends to their fingers give the distinct (but erroneous) impression that they are kin to tree frogs. Most species of this genus live in dense forest and all are nocturnal. Group size tends to be small, on the order of six to twelve individuals.

Gametes   The haploid cells produced for fusion with other haploid cells during syngamy.

Gametophyte   A haploid stage in the life cycle of plants: it is comparable to the sperm or egg of animals, but it is a free–living organism.

Gastrulation   The developmental stage where an infolding of the surface of the embryo forms the digestive tract of the adult.

Gelada baboon   Also called *Theropithecus gelada* by primatologists. This baboon is presently restricted to the mountains of central Ethiopia. It has been placed in a genus separate from the more widely recognized baboons, in part because its nostrils open to the side of its nose, rather than in front as in the rest of the primates. Gelada baboons live on the rims of gorges, where they are relatively safe from predators. Their diet is made up almost entirely of grass. Along cliff edges, their groups can number over four hundred individuals. Males have large manes and a large, red, hairless patch on their chest. Males can reach twenty kg.; females are slightly smaller, only reaching thirteen kg.

Gene   A portion of genetic material (DNA or RNA) that encodes a protein product such as an enzyme, or a structural or regulatory compound.

Gene–for–gene   Referring to the tendency for some disease organisms to evolve in step with the defensive responses of their hosts.

Genetic complementation   The condition whereby both alleles carried in the diploid state have the potential to be expressed in the phenotype of the organism. The advantages of genetic complementation are only experienced in the heterozygous condition.

Genitalia   The sex organs.

Genome   The genetic information contained in an individual, but usually used in reference to the entire set of genetic material held by a species.

Genomic   Pertaining to the genome.

Genotype   The literal reading of the genes present on the chromosomes. To be distinguished from the phenotype, which is the expression of those genes. The phenotype differs from the genotype when interactions between genes and alleles influences genetic expression.

Germinate   To sprout from a seed or spore.

Germ line   That set of cell lineages destined to become gametes. It is as opposed to the somatic line, whose members become the cells of the body not destined to become either eggs or sperm.

Gibbon   Any member of the genus *Hylobates*, which contains nine species. Closely related to the great apes, these monkeys have very long arms and legs, which they use to move gracefully through the trees. They range from China and southeast Asia to Borneo and Sumatra. All species of this genus are monogamous and live in small (three to six individuals) groups.

Gnetales   An obscure order of primitive gymnosperms.

Gorilla   The genus name for the gorilla, a very large ape (up to 350 kg.) found in west central Africa. Gorillas spend most of their time on the ground, but will climb trees. They are entirely vegetarian. Group size can reach up to thirty individuals; each group contains one to three adult males, one of which is dominant.

Great apes   Members of the primate family Pongidae, including gorillas, orangutans, and two species of chimpanzees.

Grey–cheeked mangabey   Known by the latin name *Cercocebus albigena*, these relatives of baboons are found in equatorial Africa and Tanzania. They have a specialized throat sac, which enables them to give resonating calls. They are primarily fruit–eating, but also eat buds and shoots. Their group size averages nine to eleven individuals; although the group is dominated by a single adult male, it contains other adult male members.

Group selection   Selection operating at the level of groups. To be distinguished from selection favoring high reproductive success of individuals.

Guinea baboon   *Papio papio*, to the primatologist, is found in Senegal and Gambia as well as Guinea. This species is very stocky, almost dog–like in body shape. Their coat is olive–rufous, and they lack the mane of the hamadryas baboon or the color of the mandrill. They are omnivorous savanna dwellers. They are very closely related to the other members of their genus (such as savanna baboons) and may be little more than a subspecies. Their social system is multimale/multifemale.

Gymnosperm   A member of a taxonomic grouping of plants that generally lack specialized ovarian tissue surrounding their eggs. Included in this group are ginkgos, pines, spruces, and so forth.

Gynogenetic   Said of asexual species that use sperm to trigger the development of the egg, but do not incorporate the genetic material in the developing zygote (synonymous with pseudogametic).

Haemophillus   The genus of bacterium known to cause soft chancres of the external genitalia of humans, whooping cough, and other diseases. Its name comes from the greater growth it shows when blood is added to its culture medium.

Hamadryas baboon   Known as *Papio hamadryas* to primatologists, this baboon is found only in northwest Africa, and the southwest of the Arabian Peninsula. These animals live on the savanna in large social groups of one hundred or more individuals. These large groups are subdivided into male–dominated subgroups of up to nine adult females and one male. The females are herded around by the males for all of their reproductive lives.

Hanuman langur   Known by primatologists as *Presbytis entellus*, this langur lives in extreme southern Tibet, India, Bangladesh, and Sri Lanka. It forages both on the ground and in trees for plant material, fruits, seeds, and flowers. Their social structure is highly variable depending on their environment, ranging from single–male to multimale to all–male groups. Mixed–sex groups can reach up to 125 individuals.

Haploid   Containing only one copy of each gene locus. Mammalian gametes are haploid, and then fuse to form the diploid state exhibited by the adult.

Hard–selected   Referring to situations where a particular genotype has a more nearly fixed fitness in a given environment, not influenced by social interactions (see *soft–selected*).

Heliozoans   An obscure group called the sun animacules.

Hemophiliacs   People who are unable to synthesize enough blood–clotting material to prevent themselves from bleeding uncontrollably when cut.

Herkogamous   The anatomical condition of separation of anthers and stigmas of the same flower to avoid selfing, or the waste of pollen in already self–incompatible plants.

Heterozygosity   The state of being heterozygous; also a statistical measure of the tendency for a particular gene locus or set of loci to have alternative alleles expressed in the population.

Heterozygous   Said of gene loci of diploid organisms when homologous chromosomes carry different alleles in the same cell.

Histocompatibility complex   Any of several molecules found on cell surfaces that are involved in the immune response in mammals. These proteins are very variable in mammalian populations and are responsible for tissue rejection in heart transplants.

Histogenesis   The formation and development of tissues.

Hominid   Human.

Hominoidea   The primate superfamily containing gibbons, apes, and *Homo sapiens.*

Homologous   Matching in structure, character, or origin. Said of chromosomes when they contain the same gene loci (for example, chromosome 6 from father is homologous to the chromosome 6 of the mother). Said of organs or morphological structures when they are derived from the same primitive structures (for example, the wings of bats and the forearms of people).

Homologue   Said of a structure that is homologous with another.

Homo sapiens   The species of which you are a member.

Homozygosity   The state of being homozygous.

Homozygous   Said of gene loci in diploid organisms when the homologous chromosomes carry identical alleles in the same cell.

Horizontal resistance   Resistance to a broad array of potential pathogenic agents brought about by one of just a few gene loci. As opposed to vertical resistance, where each pathogen is dealt with by a single gene (that is, more of a "gene–for–gene" defense).

Hormone   A chemical transmitted through the bloodstream that is used to signal specific tissues of the body.

Hornwort   A nonvascular plant related to mosses and liverworts.

Horsetails   Any of several species of the genus *Equisetum,* a primitive group of vascular plants.

Host   A species that supports a parasite or disease population.

Howler monkey   See *Alouatta*.

Hybridogenesis   The process where the chromosomes donated by the male are only used to produce the cells of the somatic lineage in the body of the offspring, and not the offspring's gametes. This effectively excludes the male from contributing his genes to the grandchildren.

Hybrid vigor   The increased vitality of heterozygotes caused by the expression of two alleles at a given gene locus.

Hylobates   Any gibbon, a group of monkeys closely related to the great apes, whose very long arms and legs allow them to move gracefully through the trees. They range from China and southeast Asia to Borneo and Sumatra. All species of this genus are monogamous and live in small (three to six individuals) groups.

Immunoglobins   The antibodies of the immune system.

Inbred   The state of homozygosity caused by successive brother–sister (or near relative) mating.

Inbreeder   One who engages in inbreeding.

Inbreeding   Engaging in selfing, cloning, or breeding with near relatives.

Inbreeding depression   The tendency for highly inbred offspring to be less able to survive and reproduce. Matings between close relatives tend to produce accumulations of bad genetic combinations.

Inclusive fitness   The fitness that accrues to an individual through its own reproduction and the reproduction of its relatives.

Insemination   The fusion of the sperm with the egg.

Interlocus interaction   Referring to the role of several gene loci in determining the expression of a given trait.

Invertebrate   Any member of the taxonomic grouping of organisms that lacks a backbone (for example, insects, worms, starfish, and so forth).

Japanese macaque   Called *Macaca fuscata* by specialists, this monkey is found only on the islands of Japan in nature, but is now frequently displayed in zoos. This species has long, dense fur, whiskers, and a beard. they are vegetarian and sometimes arboreal, but also feed on the ground. They will eat insects and meat when available. Groups of up to seven hundred individuals have been reported.

Juglandaceae   The plant family containing the walnut and hickory trees of the United States.

K–selected   Said of an organism whose adaptations center around survival in an environment dominated by competitive interactions as opposed to "r–selected," where the organism is a good colonist of unsaturated habitats.

Karyogamy   The process of fusion of cell nuclei of two different cells. To be distinguished from the fusion of more specialized whole cells (gametes).

!Kung San   Also called the Bushman, a hunter–gatherer group still present in Botswana today.

Lacerta   A genus of lizards common to North Africa and Europe, whose range extends nearly to the arctic circle.

Lactation   The production of milk by mammals.

Langur   Any member of the genera *Presbytis, Pygathrix,* or *Nasalis,* a collection of long–tailed monkeys from southeast Asia, the Malay Peninsula, Borneo, Sumatra, and similar areas.

Lemur   Any member of nine species of the primate family Lemuridae, all restricted to the islands of Madagascar and Comoro Islands. These animals are frequently seen in zoos. Their social organization is variable, and dominance hierarchies are not well established.

Leontopithecus   The genus containing the lion tamarin, a small (<1 kg.) long-tailed fuzz ball from old growth forests of SE Brazil.

Libido   The sexual urge.

Linkage   The tendency for alleles on the same chromosome to be passed together as a unit to the next generation. Linkage occurs because each chromosome contains more than a single gene locus. Linkage is not usually 100 percent, because crossing over during meiosis allows for fragments of DNA to be exchanged between chromosomes.

Linkage disequilibrium   The condition where particular pairs of gene loci appear to be passed together much more often than would be expected by chance and recombination alone. When there is no linkage disequilibrium, the physical structure of the chromosome might as well not exist from the standpoint of the formation of new combinations of alleles at different gene loci in offspring.

Liontailed macaque   Also known as *Macaca silenus,* from southwest peninsular India. This species has tufts of long, greyish hair along both sides of the face. Longtail macaques are found only in dense forest in the wild, where they spend 90 percent of their time in trees. Other macaques are much more terrestrial. Little is known of their social structure beyond the fact they live in multimale groups.

Liverwort   A flattened plant related to mosses and hornworts.

Locus   The physical location of a gene on a chromosome (plural: loci).

Lymph nodes   The rounded masses of tissue that assist your immune system in battling invaders by filtering the lymph and storing white blood cells.

Macaca fascicularis   See crab–eating macaque.

Macaque   Any of nineteen species of the genus *Macaca.* These animals occur in Japan, southeast Asia, Afghanistan, India, Java, Borneo, and elsewhere. Their social gatherings can be quite large (up to two hundred individuals in some species). Some have adapted quite well to living in urban situations.

Macromolecule   A very large molecule, usually referring to a large protein, carbohydrate, or lipid.

Major Histocompatibility Complex   The set of proteins that are found on the surfaces of your cells that enable your immune system to distinguish between you and an invader. In mammals, these proteins are extremely variable and account for the difficulties of transplanting tissues between different individuals (for example, kidneys, lungs, hearts).

Mandrill   *Mandrillus sphinx* is the scientific name for this species, which occurs in equatorial Africa. Mandrills are closely related to (and therefore look quite a bit like) baboons, but the bright facial coloration of this species makes them stand out. Mandrills have prominent purple and blue ridges that parallel their nose and a green upperside to their coat with yellow underwear. They feed primarily on the ground, eating fallen fruit but also looking under stones and leaves for insects and small vertebrates. The basic social structure is a single adult male surrounded by five to ten females with young. During the dry season, these small groups fuse to form gatherings of over two hundred individuals.

Marmoset   Any member of four genera: *Cebuella, Callithrix, Saguinus,* or *Callimico.* These are the smallest of the primates (often less than 1 kg. in body weight). They often have comical tufts of hair or other adornments on the head. They are found active during the day in the forests of Central America to central South America. They forage for fruit, buds, insects, and tree sap. Their social systems are very poorly known.

Mating type   The designation of one kind of cell as a potential partner for fusion with another cell of a different mating type. Thought to be a precursor to the designation of gender seen in animals.

Meiosis   The process of cell division that results in the formation of four haploid gametes from a diploid cell. The process involves DNA replication followed by two cell divisions resulting in four cells with only one copy of each gene locus.

Meiotic   Pertaining to meiosis.

Menstrual   Pertaining to the ovarian cycle.

Menstruation   The visible loss of blood by the female at the end of the ovarian cycle, when the lining of the uterus is being shed.

Methylation   The attachment of a methyl group onto a preexisting molecule. One of the biochemical changes that occurs in the formation and repair of DNA.

Microbe   A microscopic organism.

Microcosm   The world of microscopic organisms.

Midcycle   The time of ovulation in the fertility cycle of mammals.

Mitosis   The process of cell division that results in the formation of a pair of cells with the same genetic information as the parent cell from which they arise. To be distinguished from meiosis, where similar forms of chromosomal movements result in the formation of four cells with decidedly different genetic contents than the parental cell.

Mitotic   Pertaining to the process of mitosis.

Mollusks   A taxonomic grouping including the snails, slugs, clams, and so forth.

Monimiaceae   An obscure, primitive family of plants.

Monoculture   The situation where a single species (such as corn) is cultivated in monotonous plantings that do not include other crop species.

Monoecious   The condition of producing both male and female flowers or floral structures on the same individual plant.

Monogamous   Pertaining to monogamy.

Monogamy   The mating system where each sex mates with only one other individual.

Moor macaque   Also known as *Macaca maura*, from southwest Celebes, Indonesia. This species is a stocky, stump–tailed omnivore whose social groupings can exceed one hundred individuals. Smaller subgroups frequently emerge out of these larger groups and are generally headed by an older male. Younger males are tolerated in these subgroups, which range in size from five to twenty–five individuals.

Motile   Having the potential to move under its own power.

Mouse lemur   Any of three species of the genus *Microcebus,* all from Madagascar. These are small (less than 0.5 kg.), nocturnal, leaf–nest–constructing lemurs whose social system is poorly understood. Females frequently form groups of up to nine individuals, while males generally nest alone or in pairs. They feed on flowers, fruit, tree gum, and insects.

Multicellular   Having more than a single cell.

Multilocus   Having more than one gene locus involved in the expression of a trait.

Mutation   A spontaneous heritable change in the genetic material.

Myristicaceae   A plant family of South and Central America that includes nutmeg trees and other species.

Natural selection   The process whereby those entities (individuals, lineages, groups) that have the genetic tendency to successfully reproduce relatively more often tend to predominate in future generations. As opposed to artifical selection, where people decide which organisms will become the parents of future generations.

Nectaries   Plant structures that produce nectar for consumption by animals.

Neogerontology   The science of cloning.

Neutral alleles   Alleles whose relative fitness is such that they neither increase nor decrease in the population. Neutral alleles of this type may allow neutral polymorphisms to arise.

Neutral polymorphism A polymorphism maintained in a natural population because the variation between individuals is selectively neutral (that is, it does not influence their reproductive success).

Niche That set of environmental descriptors (both biotic and abiotic) that define the ecological activity space of an organism. The concept of the niche includes the habitat requirements of the organism as well as its impact on others.

Nocturnal Active at night.

Nonadditive gene interaction The situation where there is no linear relationship between the number of copies of a given allele carried in the individual (that is, the gene dose) and the intensity of expression of the trait. For example, an individual that carries AA may be three feet tall, one that is Aa would be four feet tall, and one that is aa would be five feet tall. Here the possession of an a allele gives a one–foot increment in height. This system is additive. In a nonadditive system, gene interactions influence the expression of the trait such that AA is three feet tall, Aa is three and one–half feet tall, and aa is nine feet tall.

Nonmotile Unable to move under its under power.

Normalizing selection Selection for traits of average character as opposed to selection for extreme character states (for example, selection for uniformity rather than oddity). Contrasted with directional selection, in which selection favors one of the two possible extremes.

Nucleated Said of a cell containing a nucleus.

Nucleoli A small membrane–bound structure found within the cell nucleus.

Nucleus A membrane–bound structure that surrounds and protects the genetic material of a cell (plural: nuclei).

Nurturant Providing nutrients or care.

Olive colobus monkey This species (*Colobus verus*) is found in the forests of Sierra Leone to Ghana. Their backs are olive green and their faces and underparts grey. They feed almost exclusively on leaves. They move in mixed–sex multimale groups.

Omnivorous   Eating both plant and animal material.

Oogamous   Employing a large nonmotile egg and small motile sperm to engage in sexual reproduction

Oogamy   The process of sexual reproduction using egg and sperm.

Oogenesis   The formation of eggs.

Orangutan   The Latin name for this species is *Pongo pygmaeus*, reflecting its close affiliation with other great apes. Presently they are found only on the islands of Borneo and Sumatra, where they feed primarily on figs, other fruit, insects, and bird eggs. This is a very large primate; males reach ninety kg. in the wild, with females attaining fifty kg. They move through the forest in very small groups or alone.

Organogenesis   The formation and development of organs.

Orgiastic   Resembling an orgy, or unrestrained indulgence in sexual activity.

Outbreed   Breeding with individuals that are not near relatives. Opposite of inbreed.

Outcross   Synonymous with outbreed.

Ova   Eggs (singular: ovum).

Ovarian   Pertaining to the ovaries.

Ovaries   The egg–producing organs of animals.

Overdominance   The condition where the heterozygote has a higher relative fitness than either of the homozygotes at a particular gene locus.

Ovulation   The process whereby an egg is released by the female.

Ovule   The plant structure containing the embryonic tissue that later develops into the seed.

Owl monkey   Also called night monkeys or *Aotus trivirgatus*, these monkeys range from Panama through the Amazon valley to Northern Argentina, where they live in small family groups of two to four individuals. They are nocturnal and feed on fruits, seeds, leaves, bark, flowers, tree gum, and rarely some insects or small invertebrates. They sleep during the day in hollow trees or nests constructed from accumulations of leaves. At night they forage near these day roosts. Their nocturanl vision is good, their eyes large, and their faces very owl–like. Little is known of their social organization.

Ozone A molecule composed of three atoms of oxygen that acts as a filter, blocking much of the incoming solar radiation.

Paleontologist One who studies fossils.

Paleontology The study of fossils

Pan The genus of primates that contains the two species of chimpanzees, both of which occur in the wild only in equatorial Africa.

Papio The genus name for the seven species of African baboons.

Papio cynocephalus Otherwise known as the yellow baboon from equatorial Africa. Its name comes from its yellowish coat. These baboons are omnivorous and have even been observed to capture hares. They live in open woodland and savanna in multimale groups.

Parasite Any organism that draws nutrients from another without immediately killing it.

Parthenogen Any organism that engages in parthenogenesis.

Parthenogenesis The process where offspring are genetically identical to the parent. This can arise in several ways: automixis, mitotic divisons, and so forth.

Parthenogenetic Engaging in parthenogenesis.

Patas monkey Called *Erythrocebus patas* by primatologists, this species ranges from Senegal to Ethiopia and south to Tansania. This long–tailed lanky monkey has a mustache. It feeds primarily on grasses, but also on fruit, insects, bird eggs, and so forth. It aggregates in groups of about fifteen, headed by a single, dominant male.

Pathogen Any organism that causes disease in the broad sense of the word.

Perennial Any organism having a life cycle of more than one year.

Perineal Pertaining to the area between the anus and the genitalia (synonymous with anogenital)

Phagocytotic Pertaining to cells that engage in swallowing their food or prey whole through the use of invagination of the cell membrane.

Phenols A large class of organic molecules, many of which are used as defensive chemicals by plants to protect against herbivore damage.

Phenotype The outward appearance of the organism. To be distinguished from the genetic instructions contained within the organism. For example, the genotype may encode both brown hair and baldness, but the phenotype would be bald.

Photosynthesis The biochemical synthesis of energy–rich chemical bonds through the use of light energy and carbon dioxide.

Phytopathologist One who studies plant diseases.

Pigtailed macaque *Macaca nemestrina* is the name given it by primatologists, who found it in Indochina, southeast Asia, the Malay Peninsula, Sumatra, and Borneo. It lives in dense inland forests in multimale groups of up to fifty individuals. Little is known of its social habits.

Pistillate Said of a flower containing no male structures.

Planktonic Living by floating in upper layers of water, said of many aquatic microscopic organisms and the juvenile stages of many marine vertebrates and invertebrates

Plasmid A small fragment (sometimes a loop) of DNA that is carried along with the contents of the cell, but originates independently of the host cell's DNA.

Pneumonic Pertaining to the lungs or respiratory system.

Poeciliopsis A genus of freshwater livebearing fish found in the southwestern United States, Central America and elsewhere. They typically feed on algae, plant material that falls into the water, and aquatic insects when available.

Polar body The discarded chromosomal remains of the process of meiosis in females. Rather than producing four haploid cells at the end of meiosis, as do males, females produce one large egg and discard the rest of the nuclear material that would have gone to form three other haploid cells.

Pollen The haploid gamete of plants whose function is to immediately fuse with an egg.

Pollen tube The structure that germinates from the pollen grain and carries the pollen's genetic material down to the ovule.

Pollination The process of deposition of pollen on a receptive stigma.

Polyandrous A mating system where a single female mates with several males.

Polygenic   Involving many gene loci.

Polygynous   A mating system characterized by one male mating with several different females, all of whom mate only with the single male.

Polymorphic   Exhibiting a polymorphism.

Polymorphism   The existence of variability in the outward appearance of members of the same population. Judging from the diversity of human faces, one would conclude we are a polymorphic species.

Potamopyrgus antipodorum   A freshwater–lake–inhabiting snail that has been recorded as producing offspring both parthenogenetically and sexually.

Predator   Any organism that feeds on another and in so doing kills it in a short period of time.

Prehominid   A primate ancestor of modern man.

Presbytis   The genus name for fifteen species of langurs from India, southeast Asia, the Malay Peninsula, and nearby islands.

Presbytis entellus   *See* hanuman langur.

Primate   An order of mammals including tree shrews, bushbabies, lemurs, monkeys, apes and humans.

Primatology   The study of primates.

Proceptivity   In a state of willingness or desire to copulate.

Procreation   The act of producing offspring.

Progeny   Offspring, children, children's children, and so forth.

Prokaryotae   The taxonomic classification that includes the bacteria.

Prokaryote   Organisms without internal membranes in their cells (that is, they lack, among other things, nuclei, and nuclear membranes). They are thought to be primitive to the eukaryotic state.

Promiscuous   A mating system where individuals of either sex mate with various individuals (not necessarily indiscriminately, however).

Propagule   Structure produced in plants for the purpose of dispersal and colonization of a new site, also a plant or animal colonist.

Prosimian   Referring to one of the two suborders of the primates, the Prosimii, which includes the tree shrews, lemurs, and galagos.

Protease   Any enzyme that breaks down proteins.

Protist   Any of a large nontaxonomic grouping of single–celled organisms, such as yeast, bacteria, algae, and protozoans.

Protoctist   A member of the Kingdom Protoctista.

Protoctista   The taxonomic grouping of eukaryotic aquatic or-ganisms that do not start life as an embryo (for example, giant kelp, algae, slime molds).

Protosex   The state of sexual beginnings.

Pseudogametic   Pertaining to species that employ pseudogamy.

Pseudogamy   The use of sperm to trigger the development of an egg, but where the genetic material contained in the sperm cell is not incorporated into the developing offspring, as occurs normally in sexual organisms.

Psilophytes   The earliest–known vascular plants.

Psychical   Pertaining to the mind or spirit.

Psychophysiological   Pertaining to sensations produced by a mixture of mental and cellular processes.

Pygmy chimpanzee   Called *Pan paniscus* by primatologists, these chimps live in central Zaire where they inhabit swamp forest. They are omnivorous and forage in multimale/multife-male groups of fifteen to forty individuals.

Quadrupedal   Having four legs.

Radially symmetrical   Having a wheel–like organization such that a plane in any vertical orientation divides the structure into two mirror images (for example, the flower of a daisy). See *bi-laterally symmetrical.*

Rana   The genus of frogs containing the common leopard frog.

Recessive   Said of alleles that are not expressed when carried in the heterozygous state.

Recombination   The exchange of genetic material between ho-mologous chromosomes. Homologous chromosomes often fuse at points along their length during meiosis. For example, the chromosome carrying the eye–color gene locus inherited from the father fuses with that of the maternal chromosome. Once the two chromosomes are separated during the final stages of meiosis, it is quite likely that some of the gene loci originally on the chromosome donated by the father will have exchanged po-sition with gene loci on the chromosome inherited from the mother. Recombination allows for new gene combinations on the same chromosome and breaks up linkage groups.

Red colobus monkey Any of several species of leaf–eating monkeys of the subgenus *Pilocolobus* from equatorial Africa. They live in groups of twelve to eighty–two individuals, averaging about fifty. Dominance hierarchies are well–developed, with several adult males and up to three times as many females per group.

Red Queen hypothesis The idea that organisms must continually evolve new adaptations to keep apace with the evolution of their competitors and predators. Derived from the Red Queen of *Alice in Wonderland*, who needed to run just to stay in the same place.

Redtail monkey Known to the scientific community as *Ceropithecus ascanius*, this monkey is also called the black–cheeked white-nosed monkey. It lives in Africa, ranging from Cameroon to Uganda and Zambia. It forages on fruits and plant material, but takes meat when available. Adults weight up to seven kg. They move in multimale groups of up to two hundred individuals.

Reduction divison Referring to the process of meiosis in which a diploid cell is reduced through cell division to a set of haploid cells.

Reductionist One who attempts to answer questions or solve problems by breaking the task into small pieces and trying to understand the fragments to build an understanding of the whole. Reductionist thinking has served some areas of science very well (for example, molecular biology and physics), while it has been argued that some problems (such as the organization of ecological communities, the working of the brain) can not be solved without an understanding of the whole rather than the sum of the parts.

Rhesus macaque Also called *Macaca mulatta*, this macaque ranges from Afghanistan to India, Nepal, Indochina, and south China. It is commonly used as a lab animal in the United States. In nature it feeds primarily on fruit, but also takes meat when available. Groups are multimale and can be very large (up to two hundred individuals).

RNA Ribonucleic acid; the genetic material for some viruses and other simple organisms.

Rotifer Any of a taxonomic group of freshwater microscopic invertebrates characterized by a ring of cilia that causes them to resemble rotating wheels or barrels.

Saguinus oedipus   Also called the Cottontop or Pinché Tamarin, found from southeastern Costa Rica to northwestern Columbia. This species has a prominent crest of white extending over the back of the head like a mohawk haircut. They eat fruit most often, but will take meat when availalbe. They live in small (three to nine individuals) groups, which appear to be headed by a single dominant pair.

Saimiri   The genus that contains the squirrel monkeys (*S. sciureus* and *S. oerstedii*) found from Costa Rica to Paraguay. These primarily fruit–eating omnivores are of very small body weight (less than 1 kg.). They forage in large multimale groups sometimes reaching three hundred individuals in size.

Savanna baboon   Any member of several species of the genus *Papio*, *P. cynocephalus*, *P. anubis*, and *P. ursinus*, all from equatorial Africa. Their social system varies depending on the species.

Second Law of Thermodynamics   The idea that all processes occurring spontaneously with a closed system result in an increase of total entropy.

Selection   *See* natural selection.

Selfer   One who engages in selfing.

Self–incompatibility   A fixed avoidance of inbreeding controlled by genetic markers that inhibit fertilization of eggs by pollen of near relatives or by self–pollen.

Selfing   The process of producing offspring without using genetic material from outside sources.

Senescence   The process of aging.

Senescent   Aged.

Sex   Definition varies depending on the particular usage of the author, but generally the exchange of genetic material between different lineages resulting in the origin of a novel lineage.

Sexual dimorphism   The condition in which members of different sexes are readily identified as such. Many birds, insects, and some vertebrates (people included) can be easily distinguished to sex. Sexual dimorphism is thought to arise from sexual selection.

Sexual selection   The differential reproductive success of sexual individuals (by way of mate choice, male–male competition, and so forth) that results in greater representation of the genetic material of those individuals in future generations.

Sib Shortened form of sibling, meaning brother or sister.

Sister strands The strings of genetic material contained in homologous chromosomes.

Soft–selected Said of organisms whose reproductive success is based on a social ranking scheme where the socially dominant individuals get a disproportionate amount of the resources necessary for reproduction and lower–ranked individuals get much less, regardless of the underlying genotype of the excluded individual. As opposed to hard selection, where fitness is much less dependent on social interactions.

Somatic Said of cells that form the body but do not contribute their genetic material to gametes. Somatic cells are distinguished from germ cells, which form egg and sperm in animals.

Sooty mangabey Called *Cercocebus atys* by some primatologists, its near relatives include the macaques and baboons.

Sorogena Any member of a genus of simple aquatic organisms propelled by hair–like appendages called cilia.

Spider monkey Any member of the genus *Ateles*, a group of South and Central American long–limbed monkeys commonly found in zoos. In the wild, these monkeys form multimale groups of up to öne hundred individuals.

Spirochaete A corkscrew–shaped prokaryote.

Spore A haploid propagule that will form a free–living haploid organism. To be distinguished from a gamete: a haploid cell that immediately joins another gamete to reconstruct the diploid state.

Sporophyte The diploid stage in the life cycle of a plant. Sporophytes give rise to haploid spores, which later become gametophytes.

Stamen The slender–stalked pollen–bearing structure of flowers.

Staminate Said of a flower that contains no female structures.

Staminoid A petal–like structure that serves as both a visual attractant for pollinators and as a pollen–producing structure.

Stigma The surface of the female structure on which pollen germinates and begins its migration to the ovule.

Stolon A runner or underground stem that is produced by mitotic cell division (for example, as seen in strawberries).

Stump–tailed macaque   *Macaca arctoides* in Latin, this species ranges from Assam to south China and the northern Malay Peninsula. These macaques often live in high elevation forests, where they are primarily arboreal. They feed on fruit but are also omnivorous. Groups of twenty to one hundred individuals headed by a single male, but containing several subordinate males, have been reported.

Stylar   Pertaining to the style of a flower.

Style   The filament that supports the stigma in flowers.

Sumatran long–tailed macaque   The Sumatran form of the crab–eating macaque. Once thought to be a separate species (*Macaca irus*) but recently renamed *Macaca fascicularis*. See the glossary entry for crab–eating macaque.

Symbiont   One member of a pair of organisms engaged in a symbiotic relationship.

Symbiotic   pertaining to interactions between different species, where both species benefit from the association. To be contrasted with parasitic relationships, where one benefits to the detriment of the other.

Sympatric   Said of species that live in the same geographic area. To be contrasted with allopatric.

Syngamy   The mixing of the genetic material of two cells to form a new individual.

Talapoin monkey   Called *Miopithecus talapoin* by specialists, this monkey from Gabon, Congo, southwestern Zaire, and northwestern Angola is the smallest monkey from Africa (they weigh a little over 1 kg.). It is diurnal, but little is known of its social structure.

Tamarin   Any member of a South and Central American genera *Saguinus* or *Leontopithecus*. Eleven species are included in this group. All are omnivorous and live in multimale groups of less than fifteen individuals.

Tannins   A large and mixed group of organic molecules that display a tendency to cause proteins to precipitate out of solution. Used by plants as a generalized defense against herbivores by making the herbivore's digestive enzymes unable to work in the animal's digestive tract.

Tarsier   A primitive primate of the genus *Tarsius*, which contains three species. These animals are presently restricted to the forests of Indonesia, Sumatra, Borneo, the Philippines, and the like. They are all nocturnal and feed on insects. They have a highly developed social system of monogamous pairing within larger assemblages of individuals.

Taxonomic   Pertaining to taxonomy.

Taxonomy   The study of the classification of living things, using the idea that organisms can be placed into related groups based on shared characters

Technoembryology   The study of the techniques necessary to direct the embryonic development of clones.

Testes   Synonymous with testicles (singular: testis)

Testicles   The male sex organs that produce sperm.

Tetraploid   Said of a cell containing four copies of each gene locus. To be distinguished from both the haploid and diploid states, where one or two (respectively) copies of each gene are present in each cell.

Tiller   A runner or stolon of a plant, produced through vegetative (mitotically derived) reproduction.

Transmutation   To change from one form into another.

Transposition   The insertion or rearrangement of the genetic material through enzymatic "repair" processes. A fragment of one chromosome can be excised and inserted onto the end or middle of another chromosome, or cut, inverted, and reinserted into the same chromosome.

Trimeniaceae   An obscure family of primitive plants.

Triploid   Said of a cell containing three copies of each gene locus (that is, three homologous chromosomes occur in the cell).

True–breeding   The condition where a given individual always produces offspring similar to itself for a given trait, no matter what is the state of the trait in its mate.

Unicellular   Single celled.

Vegetative reproduction   Asexual reproduction, as shown in many plants where propagules are produced mitotically.

Vertebrate   Any member of the taxonomic group of organisms that has a backbone (for example, fish, people, birds).

Vertical resistance   Resistance to one or a narrowly defined set of pathogens through the action of a single gene locus. As opposed to horizontal resistance, where many pathogens can be inhibited by the action of just a single or a very few genes.

Vervet monkey Also known as *Cercopithecus aethiops*, a native of Sudan and Ethiopia. This species is primarily arboreal, but spends considerable time on the ground looking for fruit, insects, and meat. Adults weight under nine kg. Little is known of their social structure.

Vulva The external genitalia of human females.

White–collared mangabey Called *Cerocebus torquatus*, this species ranges from Senegal to the Congo (now Zaire) and is very generalized in its habitat use. It feeds on leaves and fruits and forages in multimale groups of fourteen to twenty–three individuals.

Woolly spider monkey Also known as *Brachyteles arachnoides*, this monkey occurs in southeastern Brazil and is one of the least well–known of the primates. It is presently endangered because of its use of undisturbed tall forest, a forest type that is rapidly being converted to agricultural uses in Brazil. Bands of six to twelve individuals are sometimes seen, but sightings are sporadic. The population is rapidly declining, fewer than three thousand were know to exist in 1971, no more than one thousand exist today.

Zygote A diploid cell that forms after the fusion of gametes.